Technician's and Experimenter's Guide to Using Sun, Wind, and Water Power

Richard E. Pierson

PARKER PUBLISHING COMPANY, INC.
WEST NYACK, NEW YORK.

© 1978 by

Parker Publishing Company, Inc.
West Nyack, New York

Library of Congress Cataloging in Publication Data

Pierson, Richard E
   Technician's and experimenter's guide to using sun,
wind, and water power.

   Includes index.
   1. Solar energy. 2. Wind power. 3. Water-power.
I. Title.
TJ810.P53   621.312'14   77-10120
ISBN 0-13-898601-0

Printed in the United States of America

*I dedicate this book to my father, Karl E. Pierson,
an electronic genius who inspired me to pursue
the field of electrical engineering
and its solar energy applications.*

A WORD FROM THE AUTHOR ON
THE PRACTICAL VALUE THIS BOOK OFFERS

Many experimenters and technicians recognize the extraordinary potential in building practical solar energy systems that can provide electrical power for anything from the sign in front of their business to home or farm power. For example, Bob Landing in Pleasant Hill, California, carved his own 11½-foot, three-bladed propeller, mounted it on a 45-foot tower, and using a permanent magnet motor as a generator to charge the batteries, he powered his electric motorcycle, TV's, power tools, and ham radio transmitter. John Bennett of Denton, Texas, built a canvas-bladed English windmill and used the output from a DC motor to provide auxiliary heat in his home. The government is experimenting on a large scale. Blackwell at Sandia Labs in Albuquerque, New Mexico, has built and tested a vertical wind-driver using the Darrieus (eggbeater-type) technology to produce electricity from a one-horsepower motor.

There are many useful projects that can pay for themselves over the life of the system, and you need only to design the system, locate the parts, and build it. This can be fairly easy if you have specific how-to instructions. Although there are various books and articles on how to build specific kinds of solar systems, or lists of manufacturers currently selling components for systems, most of this information has failed to answer three key questions: (1) What are the specific design criteria? (2) What do I need to know to build my own? and (3) Can I build some of the components myself to reduce the cost? This book will give you the answers to these questions . . . and much more.

A proper solar energy electrical system will eventually cost less than the commercial electricity it replaces. Such cost balancing takes into consideration maintenance of the solar energy system (which is minimal)

3

and the rising cost of commercial electricity. Solar systems are cost "top loaded." You pay for all the electricity the unit will put out over its life when you build it. Therefore in the beginning, you should avoid those systems that are too expensive, or impractical, for the average person. For example, you could purchase a Swiss-made 6,000-watt Elektro System with 5-day battery storage as a family in Maine did for slightly over $10,000. This system has a three-bladed wind generator on a 90-foot tower and is capable of powering an entire home. However, most local zoning groups do not allow 90-foot towers in residential areas and the cost is high. It's a good system for remote locations, but certainly not for experimenting or replacing available electricity. Wind-driven systems are the least expensive available. Some commercial systems are offered by Australian Dunlite and Swiss Elecktro Systems, among others. Prices are over $3,500 for a 2000-watt system and over $2,700 for a 1200-watt system. Probably, the least expensive wind driver is the Windcharger 1222H by Winco (around $500.00), a 200-watt system complete with 10-foot tower, 6-foot propeller, and unregulated 12-volt output. You provide the batteries.

The prices are getting lower, but in many cases you can get more watts by building it yourself. To simplify matters, and reduce initial costs, a building block approach can be adopted. Starting with the most cost-efficient components, a system can be built which starts putting out electricity after about a $350 investment. The initial system would be 12-volt and will reduce your electric bill slightly. Additional capability can be added in similar increments without throwing away original components. This system can be expanded to several types of solar inputs, increased battery storage, and 120 VAC output inverters as you can afford it. I have found that the most economical free sources of power are wind, exercise, water and sun in that order. This book will tell you how to use them all in a power system at *minimum cost*.

A vast improvement in capability can be attained by close attention to power loads. For example, if you want to power 1,000 watts of incandescent lighting with a 12-hour battery capability you would need a generator and about 100 amp-hours of batteries (at 120 volts). If you replaced the lighting with fluorescent lighting of half the wattage, you could create the same amount of light and need only half the power. If you light only the book you are reading instead of the whole room, you will need ¼ of the power. If the fan blows on a small area instead of over a whole room, you might turn off the air conditioner or turn it down. An

AC-DC portable television only uses half the power on DC as compared with AC because the power supply is bypassed. These are examples of the many energy conservation guidelines in this book to make your solar power system more effective.

If you have been wanting to build your solar energy power system, this book will help you do this in the most economical way. It's all here; the design, the price, the parts list, and *detailed instructions* for efficient, economical solar power systems. With the exception of Chapter 14, all the information has been validated by actual experience. In all cases, make certain that you follow local safety and building codes for your specific area.

With the price of electricity climbing rapidly, even the thought of less expensive solar energy is extremely comforting. Solar energy is not free initially, but after several years of operation you will be able to say the electricity your system is putting out is free. I find great satisfaction during commercial power failures in having the only lights and TV on the block in operation. You can have much more than that going for you if you like. This book provides a broad range of exciting and fun projects that will pay for themselves in the long run. I urge you to join the elite group of technicians and experimenters now enjoying the benefits of their own energy producing devices.

Richard E. Pierson

ACKNOWLEDGEMENTS

I wish to thank my wife, Roberta, for her invaluable help, Wismer and Becker Contract Engineers for making their reproduction facilities available to me, and Robert G. Hoehn for his encouragement in the development of this book.

ILLUSTRATIONS AND TABLES

TABLE OF CONTENTS

1.

Private Power Means
Thrift and Independence

THREE REASONS FOR PRIVATE POWER

There are three good reasons for private power: the monthly electric bill, commercial power failures, and energy conservation.

Last month my electric bill was $41.00. This month it is $34.00. What happened to cause the $7.00 decrease this month? Two things—a reduction in lighting loads by use-oriented lighting, and the beginning of a private power system. I am now writing in the light from my own 12-volt electrical system. This entire book will be written in light produced by my private system. Should commercial power be interrupted during this chapter, I will not be bothered nor stopped in my task.

My final goal in using private power is to have light, radio, TV, a small refrigerator, a small stove, and some heat on my own electrical system. Once completed, I will be very independent of commercial power. Never again to experience a complete blackout—a very comforting thought. Also, every bit of electricity that I use on my private power system is a direct savings of electricity from the commercial power system. Of course it would take a vast number of people doing the same thing to have a noticeable effect on the commercial power system, but, every little bit helps energy conservation.

HOW THE COST OF ELECTRICITY IS CLIMBING

The best reason for building your own private power system is the rising cost of electricity. For example, in one State, the cost of electricity six years ago was ½¢ per kilowatt hour (KWH). Two years ago it was 1¢ per KWH; last year it was 2¢ per KWH. My personal cost for commercial electricity was 1.89¢ per KWH 2½ years ago; today it is 2.84¢ per KWH. That is a 50% increase in 2½ years. At that rate, it will double in 5 years! (If you wish to compute your cost per KWH just divide the kilowatt hours used into the total bill.) The cost of electricity in the United States is increasing at a *minimum* rate of 15% per year, and is expected to double in 5 to 7 years.

There are good reasons for this rate of increase. The utility companies are faced with rising expenses in order to create the needed power for their customers. For example, a standard 45-foot wooden power pole now costs about $155—a 102% increase in the past 2 years. (However, transformers of smaller size have only increased in cost about 25% over the past 4 years.) The average cost increase of material purchased by Pacific Gas and Electric was slightly under 25% last year. The utility companies cannot stay in business without passing on these costs to the customer.

Electricity rates vary throughout the United States. Figure 1-1 shows how recent residential rates varied due to location. Remember this: Regardless of what you pay now, the one thing you can be sure of is that 5 years from now it will be approximately double.

FUEL SOURCES FOR ELECTRICITY

The energy crunch will affect electricity generation. Recently figures were released showing that 26.2% of the total energy consumed in the U.S. was used in the generation of electricity. The sources of fuel for this electricity were 66.5% of all the coal consumed in the U.S. and 100% of the hydroelectric and nuclear fuel used. Yet, the top fuel sources for electricity were natural gas and petroleum. Today, electricity in the U.S. is provided by the following fuel sources:

Petroleum	43%
Natural Gas	30%
Coal	19%

LOCATION	AVG. RATE PER KWH	LOCATION	AVG. RATE PER KWH
New York, N.Y.	7.3¢	Washington, D. C.	3.2¢
Juneau, Alaska	4.8¢	Rochester, N.Y.	3.2¢
Richmond, Va.	4.2¢	Phoenix, Ariz.	3.1¢
Norfolk, Va.	4.2¢	Buffalo, N.Y.	2.9¢
Detroit, Mich.	3.9¢	Portland, Maine	2.7¢
Little Rock, Ark.	3.7¢	Cincinatti, Ohio	2.7¢
Raleigh, N.C.	3.7¢	Sheridan, Wyo.	2.6¢
Toledo, Ohio	3.6¢	Youngstown, Ohio	2.6¢
Greensville, N.C.	3.6¢	Gulfport, Miss.	2.4¢
Salt Lake City, Utah	3.6¢	Las Vegas, Nev.	2.4¢
Springfield, Mass.	3.5¢	South Bend, Ind.	2.3¢
Baltimore, Md.	3.4¢	Los Angeles, Ca.	2.3¢
Huntington, W. Va.	3.4¢	Sioux Falls, S.D.	2.3¢
Evansville, Ind.	3.3¢	Atlanta, Ga.	2.1¢
Columbia, S. C.	3.3¢	Houston, Tex.	2.0¢

FIGURE 1-1. TYPICAL RESIDENTIAL RATES
(Based on 1,000 KWHs per month summer usage)

Hydropower & Geothermal	5%
Nuclear	3%
Solar	0%

The future for sources of electricity shows an increase in hydroelectric and nuclear power plants and the beginning of solar-electric power plants. According to the Federal Energy Research and Development Administration (ERDA), solar-electric plants are expected to provide approximately 25% of our electricity by the year 2020 with no single fuel source contributing more than 25% of the total electricity used. In general, utility companies should keep up with electrical demand *and* KWH prices will steadily increase at about 20% per year. In this climate, private power will become more valuable.

COMMERCIAL POWER FAILURES

Commercial power failures came to the attention of the public several years ago in the northeastern United States when a power failure caused a domino-effect loss of electricity over a large area that involved more than one state. Since then utility companies have reviewed their interconnection of one system to another to insure that automatic equipment prevents another such occurrence. In essence, large area blackouts may be considered a thing of the past.

Brownouts may occur during peak loads. A brownout is a reduction in power output caused by a lessening of the voltage supplied by the utility company to its customers. This only occurs when the power company is having trouble meeting its peak load demand. Customers with special equipment or special transformers can compensate for lower voltage by stepping it back up; the remaining customers must use the voltage they get. Limited interruptions sometimes occur and prevent any use of electricity for short periods of time and, usually move from one location to another. (These are called rolling blackouts.)

Many areas of the country lose commercial power during storms and local equipment failures. These usually involve only a few thousand residents at a time. But . . . recently, an ice and snow storm hit the Great Lakes area leaving 600,000 people in Wisconsin and 215,000 people in Michigan without power—many for as long as a week. However, most losses of power are over within a few hours.

Weather is the primary factor in commercial power failures. Those areas of the U.S. with better prevailing weather year around enjoy more reliable power. Should you be located in an area where frequent power losses occur, you may benefit greatly from your own private power system.

Now let's take a look at what some experimenters have done to make private power systems possible.

EXAMPLES OF EXPERIMENTERS' ELECTRICAL PROJECTS

John Bennett of Denton, Texas is an engineer for the Sun Oil Company. He is an inventor and is very interested in solar energy. Living in a location where propane was his home heating fuel source, he decided to do something about its high cost. One of the cost-cutting ideas he came up with was an English windmill that provides electricity to create supplemental heat in radiant heaters in

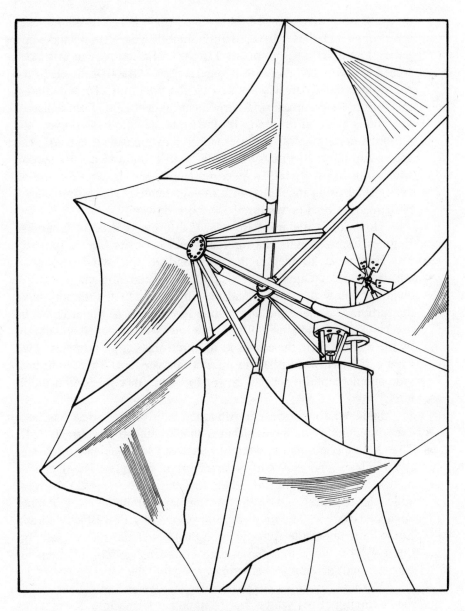

FIGURE 1-2. BENNETT'S ENGLISH WINDMILL

the walls of his den and bedroom. (See Figure 1-2.) His main blades are canvas and these drive two automobile gears connected via a pulley to a 2-HP DC motor for 115-200 VDC output. In this case, John has used a DC motor as a generator by extracting the electrical energy created when it is rotated by the windmill. To accomplish this, either field current or a permanent magnet (PM) field is necessary along with an in-line (with the armature) diode to prevent the motor from running off batteries (when they are used as a load). The speed step-up for the motor is about 50 to 1. John Bennett's system was built from inexpensive components and is an outstanding example of using ingenuity with salvaged parts. (He recommends a high speed propeller windmill for more power.)

Howard Chapman of Reseda, California is head of Chapman Engines International. His turbine wheel is on its side in an attic. The turbine is 20 feet in diameter and accepts wind through it vertically. He creates the vertical flow by allowing air to enter a chamber below the turbine and providing an exit for the air above the turbine. To make the unit respond to omni-directional winds he used louvers on the attic walls. These louvers are opened and closed by the wind to form the enclosed-on-three-sides compartments. (See Figure 1-3.) Howard plans to market this concept. This could provide supplemental electrical power to homes or commercial buildings.

Sandia Laboratories in Albuquerque, New Mexico has been experimenting with wind turbines that rotate on a vertical shaft. They built a combination wind turbine that used Darrieus (eggbeater type) blades and Savonius (barrel type) buckets. The Darrieus blades are not self-starting, therefore the Savonius buckets were added as a starter. Later the buckets were removed for a more advanced system. The units were developed by Ben Blackwell and Louis Feltz under the Aerodynamic Projects Department. (See Figure 1-4.) In 1977, this same team completed a 55-foot diameter Darrieus unit driving an induction motor into the local power grid.

Vertical rotating shaft windmills have greatly intrigued me. In the beginning of my search for cheap power I set out to develop this concept to a useful level and thus be able to include it in this book. I built a test stand where 20-foot diameter wind turbines can be rotated on a vertical shaft. I started with a modified Savonius bucket

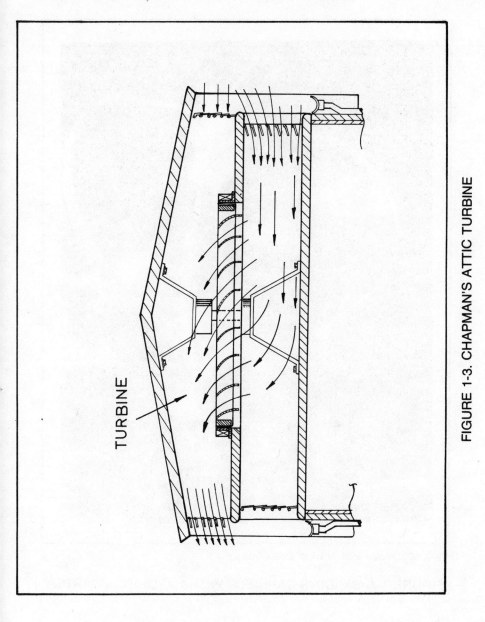

TURBINE

FIGURE 1-3. CHAPMAN'S ATTIC TURBINE

(Photo courtesy Sandia Laboratories)

FIGURE 1-4. SANDIA'S DARRIEUS WITH SAVONIUS STARTER

turbine measuring 4 feet in diameter and 9 feet high and used three 55-gallon drums that were cut in half. However, I soon found this unit to have certain deficiencies. Working closely with the wind power equation ($P = \frac{1}{2}\rho\,AV^3$), I found suitable modifications that I have tested and recommended in the chapters on wind power. (See Chapter 4 for an explanation of the wind power equation.)

Figure 1-5 shows the test stand with the first prototype windmill. The modifications that were made are reflected in the top view sketch shown in Figure 1-6. I call this turbine the Pierson Wind Turbine. In Figure 1-7 a photograph of the tested Pierson Wind Turbine is shown. The system fits atop a 2 car garage or on a barn as a second story, or as a one-story unit at a good wind location.

Experimental results for the Pierson Wind Turbine varied with configuration; upper limits depend on generator sizing. The following results were obtained in my experiments:

Configuration	Wind-Speed Begin Power out	Electric Power at 10 MPH Winds	Electric Power at 25 MPH Winds
Bare Turbine	7 MPH	36 Watts	562 Watts
Turbine with four 8 ft.-long vectoring walls	5 MPH	104 Watts	No Data (Est. 1,622 Watts)
Turbine with four 14 ft.-long vectoring walls	No Data (Est. 3 MPH)	No Data (Est. 216 watts)	No Data (Est. 3,370 Watts)

For the bare turbine (considering area equal to air pickup side of turbine only), the efficiency is approximately 20%.

Figure 1-6 shows that the wind must enter one or two of the four square funnels that feed air to the turbine. Air is exhausted out of the other square funnels. The wind velocity at the turbine is higher than the ambient wind.

As you can see, private power is there for the taking. (Although the examples have centered on wind power as a power source, many other sources are illustrated in later chapters.) Help yourself to your share and enjoy the satisfaction of independence and the local recognition that goes with it.

FIGURE 1-5. THE AUTHOR'S VERTICAL WIND TURBINE TEST STAND

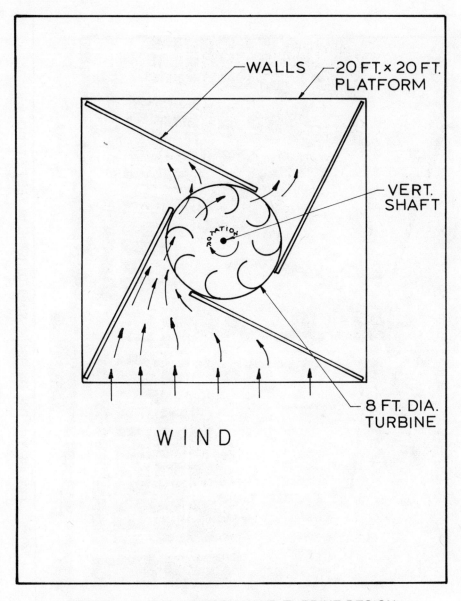

FIGURE 1-6. THE PIERSON WIND TURBINE DESIGN

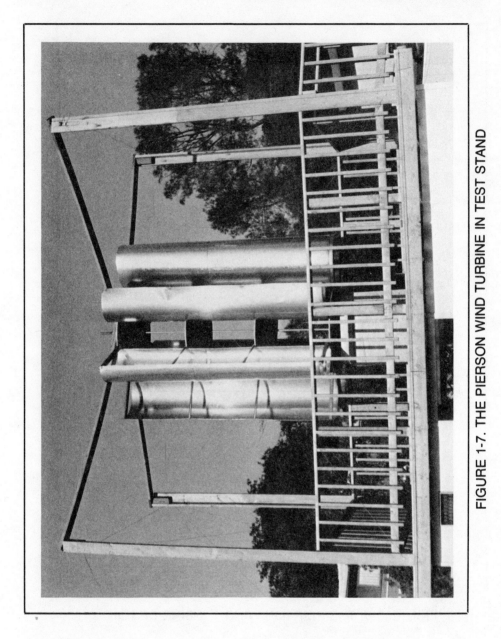

FIGURE 1-7. THE PIERSON WIND TURBINE IN TEST STAND

2.

The Economics of Free Electricity

HOW NOT TO FOOL YOURSELF ABOUT "FREE"

Is there such a thing as free electricity? If you disregard initial cost and maintenance, and use a free source of energy such as sun, wind, or water, you could say the resulting electricity is free. But let's not fool ourselves—initial cost and maintenance must be paid by someone. Whether the electrical system is the utility company's or your own, the user of the electricity always pays for it.

What is a free source of energy? Anything that heats or moves can be a free source of energy. All free sources of energy can be converted with existing devices to electricity. The most obvious heat source is the sun although not all of its energy arrives as heat. The sun's output from ultraviolet through visual light to infrared (heat) can be partially converted directly to electricity by solid-state photovoltaic devices known as "solar cells," "silicon cells" and others. The primary advantage of these is long life (plus no moving parts) while the disadvantage is the extremely high initial cost. These devices are used in the space program extensively. (Other methods of converting the sun's energy to electricity are discussed in Chapters 5 and 14.)

Other free energy sources are wind, streams, other moving objects and exercising. Wind is used as an energy source by pushing a rotor which drives a generator. Streams and other moving objects also are attached to generators to convert those movements to electricity. Physical exercise can be used for movement of generator equipment instead of wasted. The primary advantage to using moving objects as a free energy source is that they are the cheapest to convert to electricity. The primary disadvantage is that they all have moving parts which are subject to wear.

HOW TO LOWER INITIAL COSTS

With free energy sources available, you ought to be able to create a system that is more economical than commercial electricity. The only obstacles that stand in the way are initial cost and maintenance. Therefore both have been attacked with the goal being minimum cost.

There are two ways to lower the initial cost of a system:

(1)   Reduce the number of parts
(2)   Reduce the cost of the parts

Sounds simple doesn't it? But the maximum amount of ingenuity is necessary to achieve a reasonable cost level. Many initial ideas for reducing costs are found to be infeasible. It is the objective of this book to check out all cost-saving ideas known to the author and recommend the ones that are feasible.

Let's look at the functions needed to achieve use of solar power. In Figure 2-1, you can see that the oversimplification says you need only a solar source and a place to use the energy. This does not describe the hardware nor account for the periodic availability of this energy, i.e., the sun doesn't shine at night. Being careful to add only necessary components, let's proceed to identify them.

In Figure 2-2, we have added equipment to convert sun, wind, or water energy to electricity. Since the energy available from solar sources varies and is not always equal to the electrical load requirements, we need an electrical storage. These three pieces of equipment (generator, battery, and load) are the basic requirements for your system. In Figure 2-3, the possibilities for a system are shown specifically. To involve the minimum of components in your system, pick wind, water, or exercise, whichever is more readily available to you, and plan on one conversion to electricity for starting. You must have batteries in your system if you are to have electricity when the wind is not blowing, the water not running, or exercise not being done. Initial loads can be direct current (DC) such as lights,

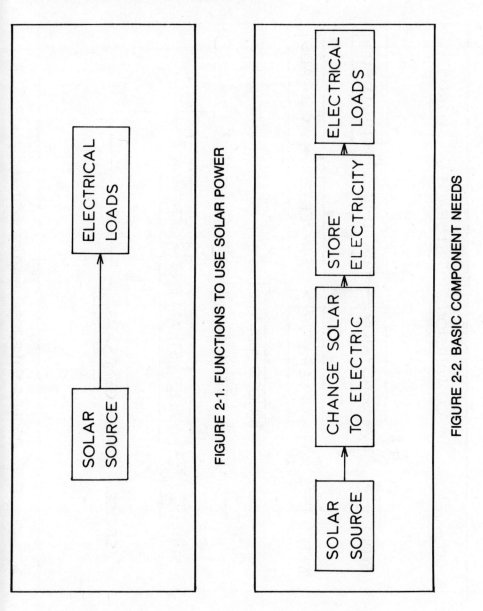

FIGURE 2-1. FUNCTIONS TO USE SOLAR POWER

FIGURE 2-2. BASIC COMPONENT NEEDS

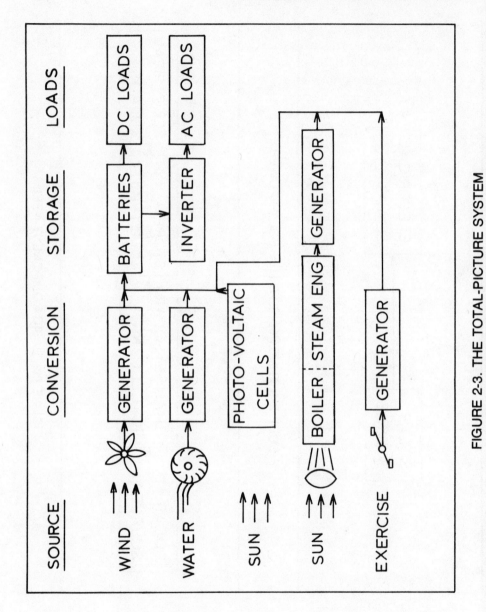

FIGURE 2-3. THE TOTAL-PICTURE SYSTEM

battery-powered TV, or special DC equipment. Later on, you can add an inverter (if you desire) to run normal alternating current (AC) loads now handled by your commercial power. In summary, the three major components you need for starting a system are a source-powered generator, a battery, and a DC load.

Now let's take a quick look at how to minimize the cost of components. The generator should be matched to the battery voltage and battery voltage should be matched to the load. In the beginning, you will have to decide on a 12-volt, 24-volt, or 120-volt DC system. The 120-volt system would match your existing lighting voltage, but would require at least 60 cells in your batteries (ten 12-volt batteries), making that an expensive choice. The 24-volt system saves most of the battery cost but you would have difficulty finding loads to match this voltage. The 12-volt system is the least expensive and generators, batteries, and loads are available at reasonable prices. Therefore, a 12-volt system is recommended for starting your system at minimum cost. (Assuming that the system voltage you choose is made up from 12-volt batteries of the same size: the higher the voltage, the larger the electrical storage capability. Although the storage capability can be matched at lower voltages by paralleling batteries, the smallest system and therefore the one with the lowest cost is a single, 12-volt battery.)

Your household 120-volt AC loads consist of your refrigerator, heater blower, air conditioner, stove (perhaps), appliances (electric frying pan, microwave oven, TV's), and lights. Which of these can be run on 12-volts DC? None, if they are not designed for it. But if they are purchased for 12-volts DC, you can run them. Therefore, I have included these loads in the initial system.

What have we decided to do with your home electricity so far? (See Figure 2-4.) Why did I limit the 12-volt DC system to *part* of your home power? Simply because the major electrical loads in your home are not easily converted to 12-volt operation. Year around, the unit that uses the most electricity in your home is the refrigerator. This is exceeded by the air conditioner or an electric heater on a seasonal basis. The refrigerator, air conditioner, and electric heater (if you have these) consume about ⅔ of your electricity. The remainder of the electricity is consumed by appliances, entertainment units, and lighting. This last group can be converted to 12-volt DC operation. (To create power for the refrigerator, air conditioning,

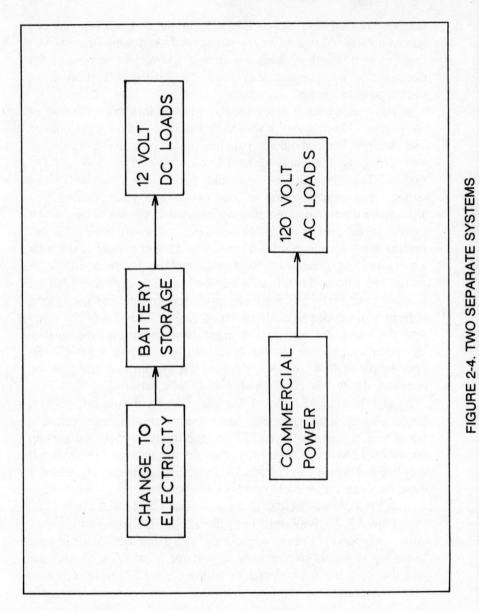

FIGURE 2-4. TWO SEPARATE SYSTEMS

and heating from your battery storage system will require a large inverter. They are expensive! Regardless, they will be discussed in Chapter 8 on using your private power. The cheapest available inverters and necessary sizes will be recommended. For those of you that consider cost to be no problem and want your own electrical power completely, be assured it can be done.)

Essentially, the 12-volt DC system will tear out a chunk of your electric bill and will replace the commercial system with a separate one.

We have now reduced the number of components that it takes to start your electric system. Commercial components designed specifically for free energy source applications are currently built in relatively small quantities. This means that they are expensive. We get around this high cost by substituting, where possible, with less expensive components that are made for other applications—such as automobile parts. A further savings in component cost can be gained by obtaining rebuilt or used parts. Some caution is necessary here, because the life of the component may be greatly reduced.

HOW TO LOWER MAINTENANCE COSTS

Lowering maintenance costs and effort is a matter of selecting the right components. I'll tell you which ones have low maintenance and how this affects the initial cost so that you will be able to judge for yourself the advantages and disadvantages. Then I will recommend which component to use.

EFFECTIVE SIZING OF PRIVATE POWER

Now what have we got? We have a private, 12-volt DC electric system that will power 12-volt lights, television, fans and other appliances. This system is capable of providing a portion of your electrical needs. Furthermore, you can start out small and add on as you can afford it. This system will provide some power when commercial electricity has failed. This system is more economical than the commercial system because the commercial system is fighting the rising cost of fuel; you use free energy. Commercial power will have to change fuel sources as fossil fuels become scarce. You won't—the wind, sun, and water will always be there.

If you build your private electric system carefully and stay within the financial boundaries that you can afford, the system will pay for itself in 6 to 10 years. Furthermore, you will have helped to take the burden from commercial electric sources, thereby leaving more fuel for other purposes.

HOW TO COMPUTE TIME TO PAY FOR ITSELF

To compute the time that it will take for a system to pay for itself, estimate how many kilowatt hours (KWH) you saved from your commercial system in a year's time. Multiply that by the current rate that you pay per KWH. For each year after the first, increase the rate 20%. When the total figure equals the cost of your system plus estimated maintenance, then you have arrived at the number of years necessary for it to pay for itself. This is only an approximation, but it does tell you where you are putting your money. After the pay-for-itself period has ended until the system life expires, your electricity is free!

Let me illustrate the point. Let's say you have built your own private 12-volt DC system. It cost you $500. It consists of a windmill driving a generator with three 12-volt batteries. The DC loads connected are replacing 1200 watts of AC lighting at an average of 3 hours per day. 3×1200 watts = 3600 watt hours per day = 3.6 KWH/day = 108 KWH/month. Now let's say you pay 5¢ per KWH for electricity. Then your initial savings on your electric bill is $5.40 per month. We'll say the maintenance on your private system is $5.00 per year average and that your electric bill rates increase at 20% per year. Now let's compare year by year.

End of Year	Your System	Saving to Electric Bill
	$500.00 initial cost	$ 5.40 per month
	5.00 maintenance	64.80 annual savings
First	$505.00	$64.80
	6.00 maintenance	20% increase = $77.76

End of Year	Your System	Saving to Electric Bill
Second	$511.00	$142.56
	8.00 maintenance	20% increase = $93.31
Third	$519.00	$235.87
	10.00 maintenance	20% increase = $111.97
Fourth	$529.00	$347.84
	12.00 maintenance	20% increase = $134.36
Fifth	$541.00	$482.20
	15.00 maintenance	20% increase = $161.23
Sixth	$556.00	$643.43
	$ 87.00 of free energy in 6th year	
	17.00 maintenance	20% increase = $193.47
	$176.00 of free energy in 7th year	

The above savings are possible for a careful builder. It is very exciting to see the results!

In this chapter, I have discussed free electricity and its real cost by assuming that you have taken a certain construction path. The low cost method is recommended to illustrate that a private system can pay for itself. Other approaches to building your own system are possible and feasible. This book will provide you with useful information, whichever system voltage or system cost you elect to work with.

3.

How to Start Your
Solar Energy System

HOW TO LIMIT YOUR INITIAL COST

The first consideration in building your private power system is cost. Available funds tend to limit how much you can do regardless of what goal you have in mind. Therefore, a logical sequence in building is to begin with a simplified system that will create some electricity with the smallest investment possible.

PLANNING TO PREVENT DISCARDING OF COMPONENTS

Begin small, and work your way up to a larger system. Every effort should be made in your beginning system to use components that can remain a part of the system when you expand. For example, you may choose as your first power source a wind turbine and generator. Perhaps the generator you really want is expensive, but could be obtained at a later date. In choosing the beginning cheaper generator, you should pick one that can be used elsewhere in your system when it is eventually replaced by the more expensive model. Perhaps its secondary use would be as a generator for a bicycle exerciser or a water wheel.

The other option is to use the exerciser bicycle as your first power source and leave it in tact so that it will be useful when you add the wind turbine or other components. In other words, plan ahead so that no initial components need to be thrown away.

THE SMALLEST SYSTEM

If you are on a rather strict budget, an exercise-driven generator and a single 12-volt battery are probably as cheap as you can go. As you can afford it, a wind turbine, water wheel, or photovoltaic devices can be added as free energy sources. More reliable and more expensive generators may be used on these systems to adjust to your budget. As more power becomes available, more batteries can be added to insure 4 days or more of reserve power. As more reserve power becomes available, more loads can be added. By following this approach to expansion of your system, you will have as much power as you can afford. Your system can fit your budget.

SELECTING THE FIRST LOADS

The best use of private power is essential to your system. To minimize cost on a 12-VDC system, you should start with lighting loads. Many inexpensive 12-volt lighting fixtures are available for campers and some of these fixtures are beautiful. Some are fluorescent which are more efficient than incandescent types at producing light. (Detailed use of these fixtures is described in Chapter 8 on using private power.) Before adding other types of loads, the maximum lighting should be added, possibly to 100% of your needs. After lighting, consider small appliances such as a portable TV, electric fan, or radios. As a last type of load, you may add inverters that will convert your battery power to 120 volts AC to run some existing household equipment.

THREE LOGICAL STEPS TO EXPANSION

A definite sequence is desirable when adding components. If you follow that sequence, you will obtain the maximum benefit from your system: 1) you must add an energy conversion unit such as a wind turbine with generator; 2) you must add a battery or batteries for storage; 3) you must add an appropriate load such as lighting. When future additions are made, you start with one again, then two and then three. Source conversion, storage, load . . . and repeat . . . for as long as your money holds out. That's the method.

How do you know how much to add each time? If your pocketbook doesn't tell you, the system will. Some unknown capability will exist when you add a source conversion unit. You then add batteries to anticipate the next added load. Then you add the next load. Now examine your system for balance. If you cannot keep the batteries sufficiently charged to handle the loads without the lights dimming on occasion, then you need more sources of electricity or improvements on the ones you have. (Keep in mind that estimates of charging time and load use are just that . . . estimates. You will have to correctly balance your system experimentally.) If the batteries stay fully charged most of the time as measured by a hydrometer, you need more loads on the system. Batteries are the storage media. To find out if they are sufficient, you can disconnect the source conversion units when batteries are fully charged and find out how many days the batteries last under normal load usage. If this number of days does not meet your personal criteria for storage, add batteries as necessary in direct proportion to the results. For example, if your batteries last 3 days under normal loads without charging inputs and you desire 6 days, you must double the quantity of batteries to double the number of days of storage. Most individuals will desire 4 to 7 days storage to cover no-wind days on a wind turbine or other possible lapses in input power.

LIMITATIONS TO SIZE OF 12-VOLT SYSTEMS

The limit to economical expansion of a 12-volt DC system is about one third of your home power. Beyond that point you should obtain higher voltages either AC or DC for major electrical loads. Usually, this means putting in inverters to obtain 120 volts AC from your batteries. The cost of these inverters varies with your needs but generally it is higher for sine wave output than for square wave. Inverter power outputs are rated in volt-amperes at some power factor or in watts. The higher the power output, the higher the cost. (I will provide information on inverters in this book, but will not provide plans using specific ones.)

The real limit to system size is initial cost. The more expensive systems tend to take longer to pay for themselves. You will have to decide which is more important: the cost of your electricity or your independence from commercial power. For example, a private system providing one-third of your electrical needs may pay for itself in six years. Whereas, a private system providing all your electrical needs *may* take 15 to 20 years to pay for itself. In unusual circumstances, the location where you want your electricity could make it extremely expensive to bring in commercial power. In such cases, a private system may pay for itself in the first year. In one case this occurred because a company needed a remote microwave relay station powered. The cost to bring in commercial electricity from 50 miles away was prohibitive. The company solved the problem by building a wind-turbine generator. By paralleling this with an array of photovoltaic cells, they kept a battery charged for station power. Most applications such as this have an average power demand that is low, making them extremely suitable to private power.

So far, general ideas on how to start your system have been given. You must make the choice of what your initial system will consist. You may have a specific application in mind such as a wind turbine, generator, battery, inverter system that will drive a neon sign in front of your business. If that is the case, should commercial power fail, you will see how dramatic a single neon sign looks on a darkened street at night.

Next, I will discuss initial cost in more detail. Rough costs will be projected here for several possible starting systems so that you may better select one that fits your budget.

THREE BUILDING BLOCKS

The three building blocks for a private system are source generation, storage, and loads. The first block to build is source generation. Which one and how much will it cost? The following listings are a very rough cost estimate that you may use in planning your budget. They assume that you have to buy all your components (scrounged resources were not available) and you do your own labor (at no cost of course).

Source Generation Blocks

• Bicycle exerciser and two $1/20$ HP 12-VDC PM Motors (used as generators) with pulley, and ammeter—approximately $100.

• Wind turbine, ½ HP 36-VDC PM Motor (as generator), 3-foot stand for vertical turbines or 30-foot mast for propeller turbines, gearing or belts and pulleys, guy wiring support, and accessories—$350.

• Water turbine with similar hardware as above for stream application—$350.

• Photovoltaic array (one square foot), 15 watt, diode output and portable stand—$500. (Lower prices for higher quantity buys.)

• Sun focusing array, steam boiler, steam engine, generator, and sun tracking equipment—in excess of $2,000.

Now that you have an idea of the cost of block one, take your pick and begin. But before proceeding with a wind turbine or water turbine, read the chapters on these equipments with regard to wind and water availability. The same advice goes for sun availability if you can afford a system that uses the sun directly. The bicycle exerciser is great for physical fitness and the better the shape you work yourself into, the lower the electric bill.

Let's assume you sighted your budget on a source generation block and are ready for the storage block. Batteries may seem pretty straightforward but quite a few options are available.

Storage Blocks

• One 12-volt, 32 amp-hour, lead-acid battery with core charge (lack of trade-in)—$25. (100 amp-hour is $45.)

• A 120-volt storage consisting of ten 12-volt batteries as above in series connection—$250. (Note: this requires a 120 to 140-volt output from your generator.) (Ten 100 amp-hour batteries—$450. When buying ten batteries, ask for a quantity discount.) (Higher amp-hour ratings have longer battery life.)

• Batteries other than lead-acid types used in automobiles and trucks may be used. These nickel-cadmium, alkaline, or lead-calcium types are usually very expensive but most are designed for 20 years life. You could put in excess of $1,000 into these more expensive batteries. Therefore, they are not considered cost-effective for private power systems.

(More information on batteries is available in Chapter 7, "Creating Electrical Storage.")

Now let's look at the loads.

Loads

- Six 10-watt 12-volt incandescent reading lamps, $5 each. Quantity may vary.
- One 30-watt, 12-volt fluorescent lamp—$25. Quantity may vary.
- Luxury best-quality spot-lamps, 50-watts or less, 12-volt—$20 each.
- Portable black and white TV set, 12″ screen, 12-volt—$100.
- Portable color TV set, 8″ screen, 12-volt—$350.
- Fan, 12-volt, single person, without guard—$10. (With guard—$15.)
- DC to AC converters will be covered in Chapter 8 on "Effective Use of Private Power."

You should now have a rough idea of what cost you can expect. However, there are still other hidden costs. Throw in $50 to $100 for wiring and another $50 to $100 for monitoring your system. Some of these costs may be negligible at first, but can add up very rapidly during expansion. So budget for some of it in the beginning.

You have my best estimate at prices you may see. I have not covered all possibilities but the above information should give you a general feel for projecting costs whichever course you may take.

HOW TO CONTROL COSTS DURING EXPANSION

As you add components or blocks to your system refer to this chapter for rough prices. Also, much detailed information appears later in the book and it should be studied before proceeding. When you add batteries, loads, or create a new power source you may find that cost is a problem. If so, consider some of these ideas for temporarily relieving initial costs.

When adding loads out of proportion, you may find a lack of battery storage and source generation; problems that financially you cannot solve instantly. In that case, buy an inexpensive battery charger such as a 6-amp unit. Occasional use of the charger will allow you to use your new loads until you are financially ready for the other needed blocks. Furthermore, if you have followed the effective use of DC loads, you will still save electricity because properly used DC lighting replaces higher wattage AC lighting. In my own personal testing program, I use a battery charger to replace the windmill during periods of modification so that loads are not effected.

With this simple approach to cost control, *know your approximate costs ahead of time,* you will not run out of money in the middle of your plans. Of course, no gold-plating has taken place in the costs presented in this chapter. Better equipment is more expensive. Know exactly what you are going to do, price the exact components, and if they fit your budget, proceed with purchase and installation. Only you can control costs. Buying without planning can be very expensive especially if for some reason you can't use the component you bought.

SOLAR ENERGY SOURCES

The key to cheap power over a long period of time is the selection of an inexpensive source of power. Let's take an abbreviated look at the solar energy and free energy available.

Solar energy is energy from the sun. This energy radiates from the sun as 7% ultraviolet, 47% visible, and 46% infrared. This energy can be described as photons striking objects at the earth's surface. The molecules, atoms, or electrons of the objects struck are raised to excited states. When they return to lower energy states, they emit long-wave radiations (heat). In general, heat from objects on the earth is the main source of solar energy. However, many side effects result from this heat.

The heat created on the earth's surface by the sun occurs in unequal hot spots because of the difference in absorption and reflection of various materials and because the earth is rotating which exposes different parts of the earth to the sun at different times. This unequal heating creates high and low pressures on the surface of the earth. Air moving from high to low pressure creates the winds. The wind is therefore indirectly caused by the sun and the earth's rotation.

Also, the heat from the sun causes evaporation of water on the earth's surface. This water vapor forms clouds. With appropriate changes in temperature, pressure, and density of the water vapor, it falls as rain, snow, sleet, or hail. The water that falls on high ground runs as streams and rivers to oceans or lakes. This movement of water then is indirectly caused by the sun.

OTHER FREE ENERGY SOURCES

Directly or indirectly, the sun creates heat, air movement, and water movement. Although most of these are unpredictable in an exact way, past records show average sunshine, rainfall and winds for any given area. These thoughts have given me the clue to other free energy sources and there are many besides the sun, but none so powerful. Let's define free energy sources as *anything that heats or moves*! Let's check the definition. The sun heats, the wind moves, streams and river move. But what of other sources of free energy? Hot springs heat (water or steam), waves move (a combination of winds and tides). The tides are caused by the gravitational pull of the moon and therefore are not solar, but are a source of energy.

There really is no limit to free energy sources. I will list some and briefly tell how they can be harnessed.

People movement: Exercise bicycle driving a generator; weightlifting driving a ratcheted generator instead of fighting gravity with weights.

One-way elevator: Two cars pulleyed together, one goes up while the other goes down. The passengers only ride down. This raises the other car and the downward speed is controlled by driving a generator. Passengers must climb steps to go up. This could also be applied to a dumb waiter where up requires a hand line.

Heat escaping from clothes dryer vent: It is moist and hot and usually dumped outside. (Although air moves in this vent any use of it will cause excess backpressure on your dryer.) The heat can be fed through a heat exchanger with blower to extract heat for winter heating.

These are just some ideas to get you thinking. They all involve *heat* or *movement*! They are not all solar (from the sun). Keep your eyes open for suitable free energy sources that may be available in your area. Remember, if it *heats* or *moves* take a close look to see if you can harness any energy going to waste. That moving object doesn't have to be solid either, it can be liquid or gas also.

Although many free energy sources exist, the primary ones are:

● Direct sun radiation collected in solar heat collectors or converted directly to electricity by photovoltaic devices.

● Wind harnessed by wind turbine with generator.

● Moving water or water at a height harnessed by a water wheel or hydroelectric plant.

Of the primary free energy sources, the sun's direct radiation (insolation) is the most abundant resource in the United States, and unfortunately it is the most expensive to harness when creating electricity. Wind and water movements are the cheapest to harness if they are available in suitable quantities in your area.

Keep in mind that a very small, free energy source available on a continuous basis can create a lot of total energy. So if you have a very small, continuous stream or a hot spring, or other similar source, a great amount of energy is available for you to harness.

In summary, check your resources for free energy. Then use the one with the most energy that is within your budget. Or use more than one (if you have them) running in parallel for higher total energy.

4.

Selecting a Wind System

Wind is a primary source of electrical solar energy. You need only to determine if sufficient wind is available and which type of wind turbine you will build. The information to determine which type, how big, how costly, how efficient, how beautiful, and how powerful your windmill will be is presented in this chapter.

WHY WIND IS THE FIRST PRIORITY

Wind is the cheapest source of electricity. Therefore, it should be your first priority for obtaining solar power. Only rare exceptions to this rule occur—for those who have average winds of less than 7 MPH due to location. To assist you in determining the wind available in your area, the following monthly and annual wind maps are shown for prevailing direction and mean speed in the United States*. (See Figures 4-1 through 4-13.) Measured locations in each state are shown as well as in Puerto Rico. These locations must be used with caution because of the effects of local topography particularly in mountainous terrain. The maximum wind record is also shown. (See Figure 4-14.) Where practical, your wind machine should be designed to withstand the maximum wind. Tornado and hurricane force winds are not practical to design against. However, for forecastable high winds such as hurricanes or tornadoes, you could design your tower to be hinged to a lay down position for tie-down purposes.

*This information was obtained from the U.S. Department of Commerce's National Climatic Center in Asheville, N.C.

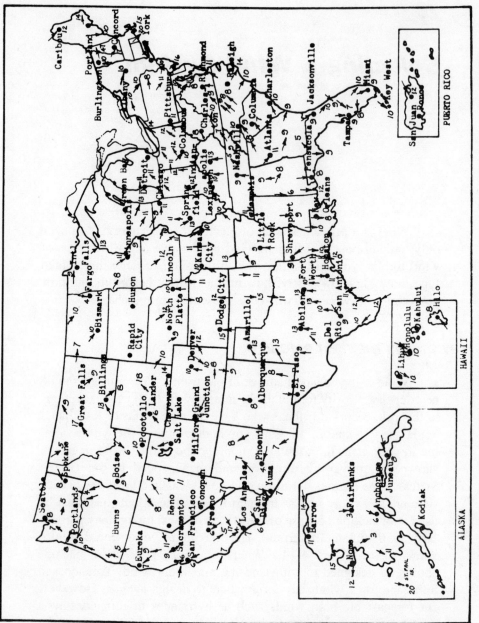

FIGURE 4-1. PREVAILING DIRECTION AND MEAN SPEED OF WIND—JANUARY

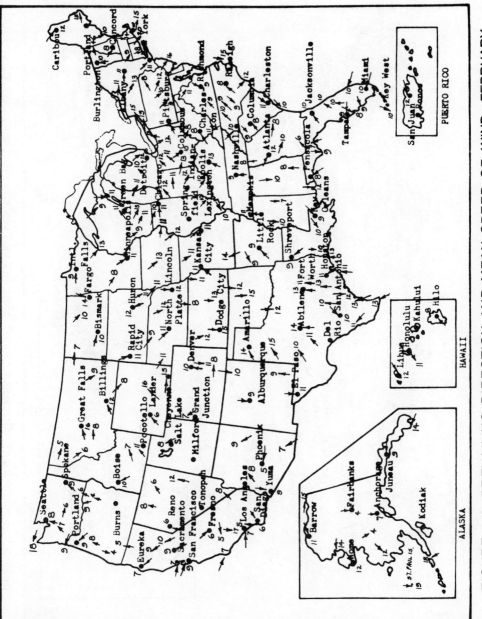

FIGURE 4-2. PREVAILING DIRECTION AND MEAN SPEED OF WIND—FEBRUARY

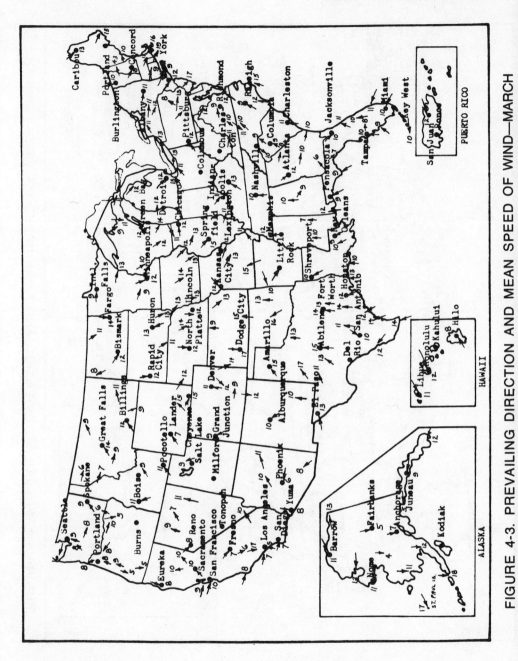

FIGURE 4-3. PREVAILING DIRECTION AND MEAN SPEED OF WIND—MARCH

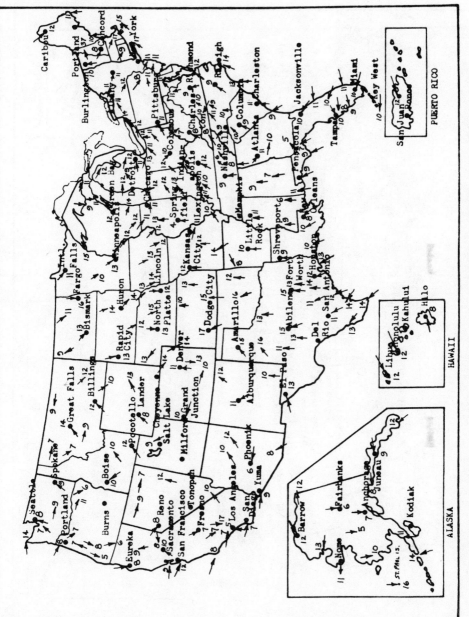

FIGURE 4-4. PREVAILING DIRECTION AND MEAN SPEED OF WIND—APRIL

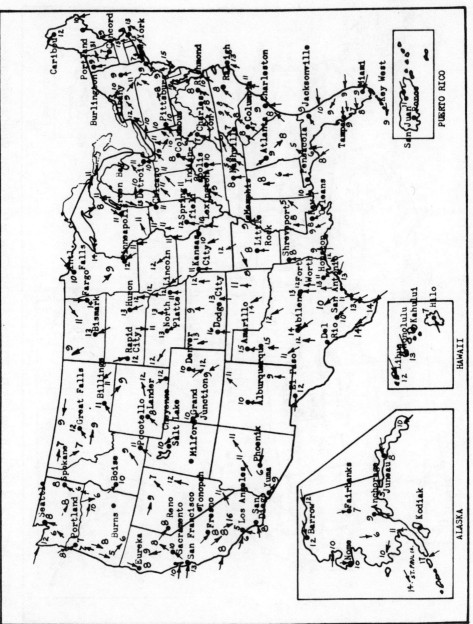

FIGURE 4-5. PREVAILING DIRECTION AND MEAN SPEED OF WIND—MAY

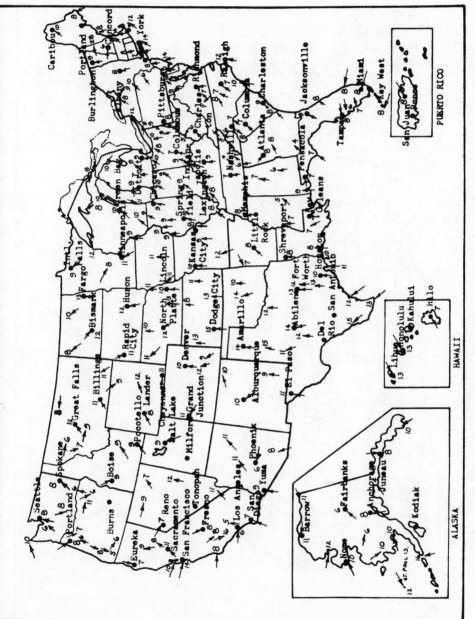

FIGURE 4-6. PREVAILING DIRECTION AND MEAN SPEED OF WIND—JUNE

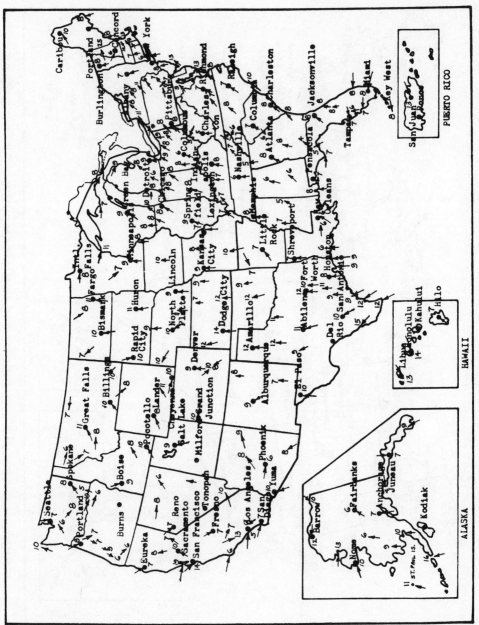

FIGURE 4-7. PREVAILING DIRECTION AND MEAN SPEED OF WIND—JULY

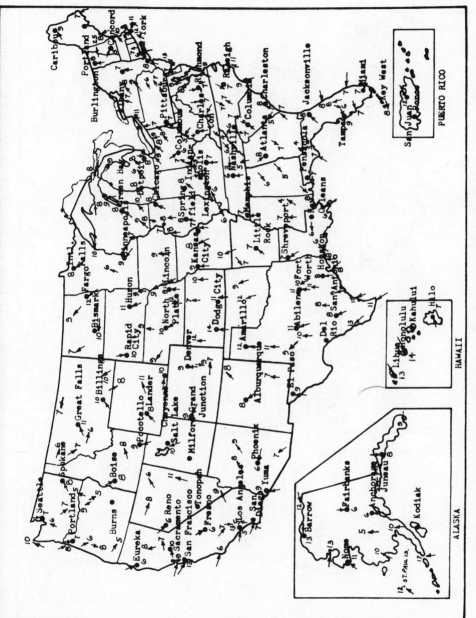

FIGURE 4-8. PREVAILING DIRECTION AND MEAN SPEED OF WIND—AUGUST

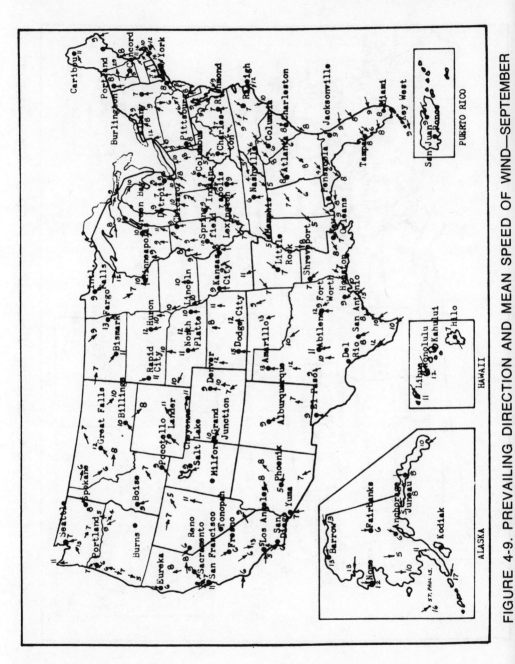

FIGURE 4-9. PREVAILING DIRECTION AND MEAN SPEED OF WIND—SEPTEMBER

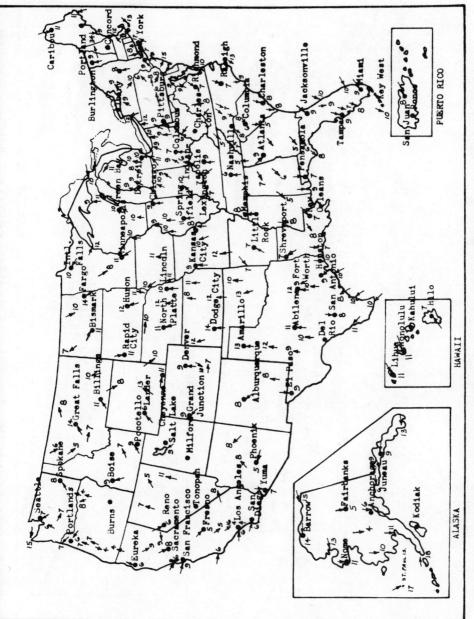

FIGURE 4-10. PREVAILING DIRECTION AND MEAN SPEED OF WIND—OCTOBER

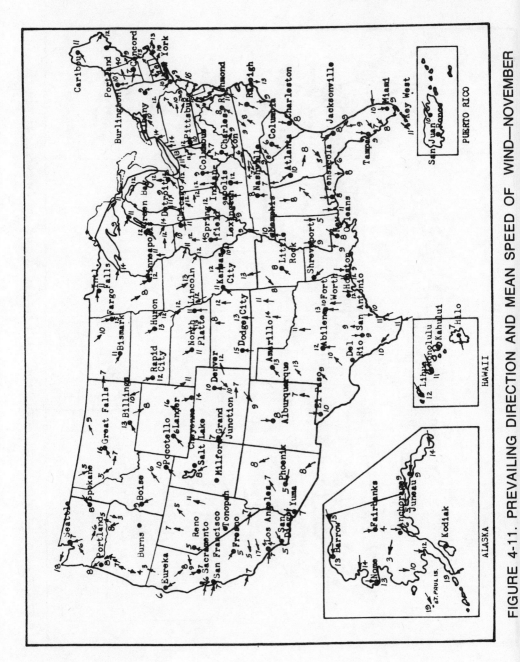

FIGURE 4-11. PREVAILING DIRECTION AND MEAN SPEED OF WIND—NOVEMBER

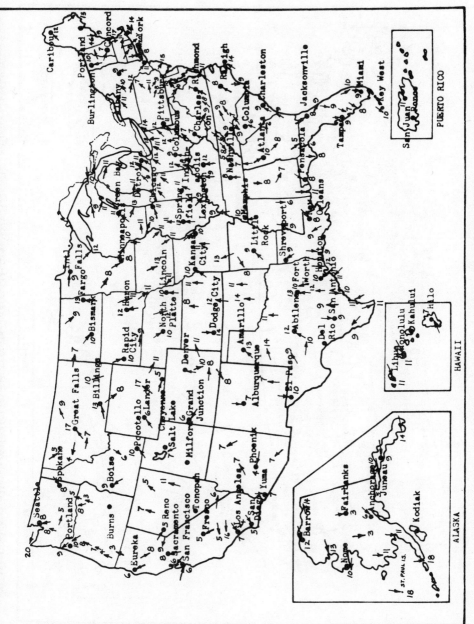

FIGURE 4-12. PREVAILING DIRECTION AND MEAN SPEED OF WIND—DECEMBER

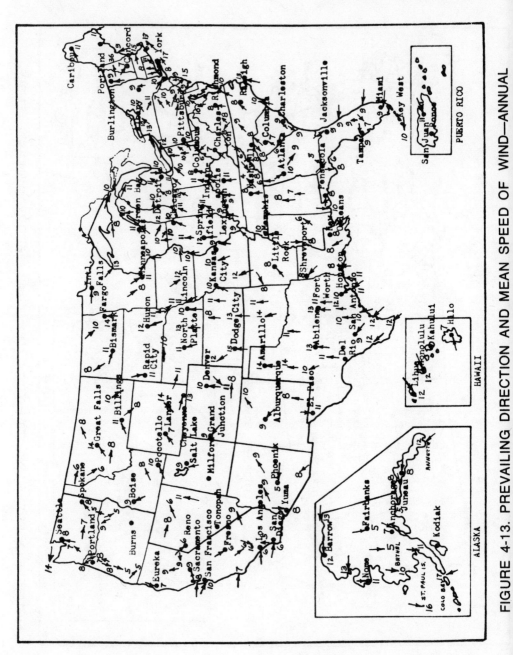

FIGURE 4-13. PREVAILING DIRECTION AND MEAN SPEED OF WIND—ANNUAL

FASTEST MPH AND DIRECTION OF WIND

LOCATION	JAN	FEB	MAR	APR	MAY	JUN	JUL	AUG	SEP	OCT	NOV	DEC	HIGHEST RECORDED FROM	YEARS RECORDED
ALABAMA														
BIRMINGHAM	49	59	65	56	65	56	57	50	50	43	52	45	65 SW	50
MOBILE	45	44	48	56	44	47	81	49	75	87	47	42	87 E	28
MONTGOMERY	47	47	60	54	50	51	51	43	40	40	46	46	60 SW	50
ALASKA														
ANCHORAGE	60	62	49	66	31	33	32	45	49	59	66	56	66 NE	10
BARROW	56	58	48	52	43	38	56	47	56	51	63	70	70 W	30
FAIRBANKS	41	57	60	35	42	40	50	34	49	50	50	45	60 SW	11
JUNEAU	68	66	49	57	47	45	32	33	48	61	59	50	68 SE	15
ARIZONA														
PHOENIX	41	49	50	45	59	46	71	60	75	48	45	68	75 SW	24
PRESCOTT	47	65	65	65	60	54	54	44	56	61	72	52	72 SW	17
TUCSON	40	59	41	46	40	50	54	54	54	47	55	44	59 E	14
YUMA	36	42	56	47	38	41	52	54	42	39	47	47	56 NW	51
ARKANSAS														
FORT SMITH	42	56	41	42	57	58	37	44	52	54	56	45	58 N	16
LITTLE ROCK	44	46	56	65	61	60	56	54	50	58	49	42	65 NW	20
CALIFORNIA														
BAKERSFIELD	47	50	35	60	40	37	28	30	32	32	45	40	60 NW	18
EUREKA	54	48	48	49	40	39	35	34	44	43	43	56	56 S	51
FRESNO	32	38	31	36	38	34	25	31	29	40	27	43	43 NW	12
LOS ANGELES	49	40	40	40	39	32	21	24	27	43	42	44	49 N	21
RED BLUFF	59	61	63	50	40	38	34	30	43	56	47	60	63 SE	17

FIGURE 4-14. FASTEST MPH AND DIRECTION OF WIND

FASTEST MPH AND DIRECTION OF WIND (Continued)

LOCATION	JAN	FEB	MAR	APR	MAY	JUN	JUL	AUG	SEP	OCT	NOV	DEC	HIGHEST RECORDED FROM	YEARS RECORDED
SACRAMENTO	60	58	66	45	40	47	36	38	37	68	70	70	70 SE	85
SAN DIEGO	39	35	46	37	27	26	18	23	25	31	51	34	51 SE	18
SAN FRANCISCO	60	62	52	48	56	48	50	45	48	51	55	53	62 SW	24
COLORADO														
DENVER	51	50	53	56	65	56	57	47	50	47	58	51	65 W	89
GRAND JUNCTION	54	50	65	59	61	66	56	56	61	61	56	48	66 S	64
PUEBLO	80	56	68	72	66	63	61	54	56	65	61	57	80 W	21
CONNECTICUT														
HARTFORD	51	50	57	47	41	52	42	56	62	57	70	54	70 E	45
NEW HAVEN	38	38	42	40	31	32	33	38	41	44	57	45	57 E	18
DISTRICT OF COLUMBIA														
WASHINGTON	56	57	60	56	48	57	54	49	56	78	60	62	78 SE	13
WASHINGTON U.	47	47	56	50	47	47	54	49	62	54	47	40	62 SE	87
FLORIDA														
APALACHICOLA	48	40	54	51	47	47	63	59	67	56	47	42	67 E	29
DATONA BEACH	38	40	40	46	32	35	40	50	45	59	29	40	59 SE	12
JACKSONVILLE	51	76	66	54	57	59	53	51	76	72	60	54	76 W, SE	90
KEY WEST	71	63	52	50	51	48	52	56	122	66	76	49	122 NW	50
MIAMI	50	68	53	70	48	48	53	62	132	122	94	52	132 E	51
PENSACOLA	35	42	49	41	40	33	35	57	91	32	35	36	91 SE	22
TAMPA	47	59	56	50	56	56	62	57	84	68	40	51	84 SE	51
GEORGIA														
ATLANTA	54	59	66	68	59	70	56	49	49	47	46	63	70 NE	50

AUGUSTA	46	43	37	41	46	43	48	46	44	37	40	39	48 -	40
MACON	47	60	60	57	65	56	59	70	43	36	54	38	70 S	62
SAVANNAH	57	62	60	64	49	66	54	90	56	85	57	51	90 N	51
HAWAII														
HONOLULU	67	63	59	40	35	39	34	52	35	41	65	59	67 SW	27
LIHUE	36	34	36	29	49	30	31	73	30	32	43	58	73 NE	11
IDAHO														
BOISE	50	56	52	50	50	50	61	50	50	56	57	56	61 W	22
POCATELLO	61	54	72	61	61	50	47	56	57	54	67	56	72 W	49
IDAHO FALLS	39	36	51	39	41	36	35	33	38	44	40	43	51 WSW	11
ILLINOIS														
CAIRO	56	56	59	59	62	52	59	56	37	40	50	43	62 N	71
CHICAGO	58	87	76	69	69	62	69	62	69	62	76	66	87 NE	90
MOLINE	56	58	66	73	68	77	65	61	66	56	60	61	77 SW	33
PEORIA	54	52	56	66	61	66	75	65	60	60	56	58	75 NW	50
SPRINGFIELD	65	63	66	56	58	75	59	58	52	44	68	66	75 SW	83
INDIANA														
EVANSVILLE	51	76	62	61	56	113	60	80	51	41	62	54	113 S	50
FORT WAYNE	59	52	65	60	54	65	61	47	56	41	57	52	65 SE	50
INDIANAPOLIS	90	65	68	60	68	111	72	68	62	56	59	53	111 NW	90
SOUTH BEND	42	47	47	42	40	50	45	63	35	38	45	43	63 NW	11
IOWA														
BURLINGTON	49	46	56	73	68	70	56	73	45	63	47	72	73 N	20
DES MOINES	66	56	70	63	71	76	66	60	55	56	72	61	76 NW	30
DUBUQUE	31	30	34	32	32	47	44	36	35	30	34	34	47 NW	76
SIOUX CITY	59	62	69	68	80	91	61	91	66	70	59	56	91 SW	50
KANSAS														
CONCORDIA	37	44	54	47	40	65	50	47	34	41	47	40	65 W	50

FIGURE 4-14. Continued

FASTEST MPH AND DIRECTION OF WIND (Continued)

LOCATION	JAN	FEB	MAR	APR	MAY	JUN	JUL	AUG	SEP	OCT	NOV	DEC	HIGHEST RECORDED FROM	YEARS RECORDED
DODGE CITY	56	72	74	70	71	73	78	72	66	63	59	57	78 SW	18
TOPEKA	43	50	66	63	67	57	81	57	57	63	49	61	81 N	50
WITCHITA	51	56	82	68	70	73	100	59	59	52	72	52	100 N	73
KENTUCKY														
LEXINGTON	52	47	53	50	50	47	49	56	53	40	52	47	56 S	45
LOUISVILLE	53	62	62	66	68	66	57	63	57	50	60	61	68 NW	49
LOUISIANA														
LAKE CHARLES	46	42	55	46	50	69	45	43	50	42	46	50	69 SE	21
NEW ORLEANS	41	44	43	35	43	40	37	41	98	31	28	31	98 SE	47
SHREVEPORT	47	45	57	52	47	50	46	52	59	47	50	52	59 E	82
MAINE														
CARIBOU	76	52	63	46	53	50	66	45	45	48	48	55	76 NW	8
PORTLAND	50	58	76	57	64	48	51	69	62	47	76	62	76 NE	62
MARYLAND														
BALTIMORE	56	65	66	65	57	61	61	60	54	49	57	56	66 W	39
MASSACHUSETTS														
BLUE HILL	72	74	69	57	59	76	67	93	121	63	70	82	121 S	53
BOSTON	66	58	73	63	55	46	52	52	87	63	80	73	87 S	43
NANTUCKET	69	71	91	61	52	68	62	79	73	69	70	62	91 E	75
WORCESTER	60	76	76	54	48	39	43	34	34	43	54	51	76 WNW	4
MICHIGAN														
ALPENA	56	50	56	52	47	50	49	52	49	50	61	49	61 SW	45
DETROIT	57	49	68	56	61	56	77	50	52	56	66	59	77 NW	28
FLINT	76	65	63	67	81	68	44	60	46	52	56	46	81 NW	19

Location													Peak	
GRAND RAPIDS	72	65	69	79	73	68	62	51	65	69	80	68	80 SW	58
LANSING	46	41	38	38	49	38	42	29	37	44	50	33	50 SW	22
MARQUETTE	52	56	57	47	91	59	61	52	45	66	50	57	91 S	49
MINNESOTA														
DULUTH	65	67	75	75	61	59	72	68	60	61	68	72	75 NE	35
INTERN'L FALLS	32	35	42	52	52	36	46	36	35	29	35	32	52 SW	9
MINNEAPOLIS	47	52	56	58	56	63	92	57	50	73	60	52	92 W	50
MISSISSIPPI														
JACKSON	56	45	68	49	56	52	54	51	54	39	51	59	68 S	14
MERIDIAN	42	41	47	44	40	40	40	36	32	32	31	38	47 -	55
VICKSBURG	38	47	40	56	49	38	46	41	47	35	47	49	56 W	50
MISSOURI														
COLUMBIA	56	45	56	57	58	58	61	56	63	49	49	56	63 NW	50
KANSAS	66	49	70	61	68	68	70	72	50	56	56	57	72 NW	71
ST. JOSEPH	46	46	64	61	63	60	52	57	43	42	56	59	64 S	48
ST. LOUIS U.	59	57	82	61	61	60	56	60	73	62	65	56	82 SW	45
SPRINGFIELD	55	59	65	60	66	57	56	59	54	49	59	55	66 W	16
MONTANA														
BILLINGS	66	68	61	72	68	65	73	66	61	68	63	66	73 N	18
GREAT FALLS	65	72	73	70	65	70	73	71	73	73	73	82	82 SW	18
HELENA	73	73	61	52	62	59	60	65	54	62	59	59	73 W	50
MISSOULA	52	53	53	51	57	51	72	58	41	51	42	56	72 SE	26
NEBRASKA														
LINCOLN U.	49	56	63	56	60	66	65	62	50	44	52	54	66 NW	50
NORTH PLATTE	57	68	72	63	70	72	60	56	53	72	64	59	72 N	50
OMAHA	60	59	73	65	73	72	109	66	51	59	58	52	109 N	89
SCOTTSBLUFF	40	60	48	46	80	80	52	41	46	44	52	47	80 WNW	10

FIGURE 4-14. Continued

FASTEST MPH AND DIRECTION OF WIND (Continued)

LOCATION	JAN	FEB	MAR	APR	MAY	JUN	JUL	AUG	SEP	OCT	NOV	DEC	HIGHEST RECORDED FROM	YEARS RECORDED
NEVADA														
ELY	66	56	59	59	74	63	50	57	57	65	51	61	74 S	18
RENO	66	66	68	66	63	66	43	57	48	66	59	72	72 S	56
WINNEMUCCA	56	59	63	49	61	53	56	51	57	54	59	69	69 E	46
NEW HAMPSHIRE														
CONCORD	43	42	71	52	48	38	37	56	61	39	72	51	72 NE	50
MT. WASHINGTON	170	144	180	231	164	136	110	142	157	161	160	175	231 SE	29
NEW JERSEY														
ATLANTIC CITY	80	77	72	80	59	57	61	76	91	72	72	70	91 NE	37
NEWARK	46	50	55	52	40	55	45	46	43	48	82	42	82 E	10
TRENTON	62	60	60	54	56	57	73	47	56	60	64	56	73 NE	49
NEW MEXICO														
ALBUQUERQUE	61	68	80	72	72	82	68	61	62	66	57	90	90 SE	30
ROSWELL	67	70	66	75	72	66	66	72	54	66	65	72	75 W	57
NEW YORK														
ALBANY	57	71	55	49	50	45	43	38	48	45	70	54	71 NW	50
BINGHAMTON	59	66	61	52	54	60	58	58	42	72	57	59	72 S	18
BUFFALO	91	73	84	73	63	73	62	60	68	71	76	85	91 SW	50
NEW YORK CITY	76	91	91	95	74	94	95	74	99	113	87	91	113 SE	46
ROCHESTER	73	66	65	59	63	61	56	59	59	65	59	57	73 W	51
SYRACUSE	56	61	59	57	49	59	52	49	47	63	59	69	69 SW	50
NORTH CAROLINA														
ASHEVILLE	52	43	44	47	49	49	40	34	43	44	40	42	52 -	50
CAPE HATTERAS	76	68	68	61	60	35	77	77	110	56	65	56	110 W	50

CHARLOTTE	57	54	47	53	48	57	50	54	47	50	47	57	57 SW,N,W,NE	50
RALEIGH	50	66	59	56	65	50	47	50	38	52	47	52	66 SW	41
WINSTON-SALEM	40	46	45	43	48	29	48	46	42	42	42	43	48 NW,SSW	8
NORTH DAKOTA														
BISMARK	70	54	65	63	66	66	72	72	66	61	67	61	72 W	87
DEVILS LAKE	41	54	54	47	52	50	56	47	46	47	57	42	57 N	57
FARGO	57	56	56	65	72	115	60	71	88	57	66	58	115 NW	41
OHIO														
CINCINNATI	49	49	49	47	36	40	43	38	38	35	47	41	49 SW	40
COLUMBUS	63	58	68	76	56	62	84	78	51	60	61	58	84 NW	59
DAYTON	60	72	75	72	60	78	74	70	65	56	68	70	78 NW	47
SANDUSKY	56	64	65	75	48	77	69	63	52	54	68	56	77 NW	84
TOLEDO	66	69	87	72	58	56	76	76	62	57	76	69	87 SW	50
OKLAHOMA														
OKLAHOMA CITY	63	61	61	75	72	87	73	56	54	65	66	56	87 NNW	47
TULSA	55	50	51	70	75	65	56	56	48	45	56	56	75 SW	19
OREGON														
PORTLAND	54	61	50	60	42	40	31	29	38	50	56	57	61 SW	50
ROSEBURG	34	38	27	29	22	22	25	25	25	33	31	31	38 SW	9
PENNSYLVANIA														
ERIE	56	57	60	56	51	47	51	58	56	51	60	53	60 SW,SE	42
HARRISBURG	47	60	68	56	46	46	47	45	40	50	58	61	68 W	50
PHILADELPHIA	62	59	61	54	73	73	88	67	49	73	60	47	88 N	50
PITTSBURGH	67	58	72	60	73	58	64	57	54	56	49	60	73 NW	79
READING	84	79	65	80	95	62	66	56	58	72	69	64	95 NW	49
SCRANTON	42	60	42	47	40	43	56	50	42	50	45	49	60 W	17
RHODE ISLAND														
BLOCK ISLAND	69	66	90	72	72	72	51	82	91	65	90	77	91 SE	70

FIGURE 4-14. Continued

FASTEST MPF AND DIRECTION OF WIND (Continued)

LOCATION	JAN	FEB	MAR	APR	MAY	JUN	JUL	AUG	SEP	OCT	NOV	DEC	HIGHEST RECORDED FROM	YEARS RECORDED
PROVIDENCE	70	63	69	65	56	60	56	49	95	60	60	85	95 SW	49
SOUTH CAROLINA														
CHARLESTON	61	52	72	65	68	54	60	73	76	56	49	73	76 -	47
COLUMBIA	54	47	54	47	47	41	51	49	35	38	43	44	54 SW	41
GREENVILLE	73	56	63	66	65	52	60	70	47	70	52	50	79 N	18
SOUTH DAKOTA														
HURON	57	56	68	73	70	65	77	70	64	72	73	56	77 NW	29
RAPID CITY	73	75	72	72	67	72	70	72	73	72	59	65	75 NW	50
TENNESSEE														
CHATTANOOGA	59	63	82	57	63	67	48	62	57	35	45	46	82 W	83
KNOXVILLE	60	61	61	71	59	65	73	50	56	42	43	52	73 SW	50
MEMPHIS	56	51	54	51	57	51	54	47	51	47	43	56	57 -	50
NASHVILLE	56	57	70	61	57	73	59	49	47	51	58	47	73 NW	50
TEXAS														
ABILENE	52	60	71	61	73	109	63	51	55	49	50	56	109 NW	49
AMARILLO	62	70	72	74	84	65	66	65	68	68	56	62	84 SW	70
AUSTIN	44	57	48	44	53	50	43	49	43	52	48	49	59 N	35
CORPUS CHRISTI	59	50	56	54	69	55	56	100	110	47	62	47	110 NE	75
DALLAS	66	61	59	58	65	65	77	56	48	61	56	47	77 N	48
EL PASO	61	69	70	66	70	68	65	63	58	58	57	66	70 NW	83
FORT WORTH	61	61	59	53	68	57	56	49	49	56	53	53	68 W	38
HOUSTON	50	53	84	63	60	54	73	80	66	70	46	49	84 NW	52
SAN ANTONIO	56	56	63	61	73	59	54	74	49	57	57	51	74 NE	50

LOCATION	JAN	FEB	MAR	APR	MAY	JUN	JUL	AUG	SEP	OCT	NOV	DEC	FROM	RECORDED
UTAH														
SALT LAKE CITY	52	56	71	56	57	54	53	58	61	67	63	54	71 NW	50
VERMONT														
BURLINGTON	62	57	66	63	49	50	42	54	56	70	72	57	72 SE	55
VIRGINIA														
LYNCHBURG	57	48	48	62	56	47	47	46	47	62	46	46	62 W	50
NORFOLK	62	62	63	62	70	80	63	70	75	78	53	62	80 W	50
RICHMOND	62	51	51	62	56	53	56	50	45	68	47	46	68 SE	50
WASHINGTON														
SEATTLE	63	64	60	65	45	54	38	35	55	63	57	60	65 S	28
SPOKANE	56	54	56	43	49	47	46	38	38	56	54	56	56 W,SW	49
TATOOSH ISLAND	87	84	91	73	66	70	53	59	68	73	94	85	94 S	59
WALLA WALLA	42	47	62	41	32	37	36	26	51	54	67	47	67 SW	46
YAKIMA	32	31	34	34	30	30	29	25	35	29	29	32	35 SW	18
WEST VIRGINIA														
PARKERSBURG	45	45	47	47	43	49	62	37	61	38	66	35	66 NW	73
WISCONSIN														
GREEN BAY	61	66	68	57	109	73	70	56	66	66	67	61	109 SW	49
LA CROSSE	35	36	40	50	58	60	36	46	36	38	46	43	60 NNW	10
MADISON	68	57	70	73	77	59	72	47	52	73	56	65	77 SW	50
MILWAUKEE	62	58	73	66	72	66	59	56	62	60	72	62	73 SW	50
WYOMING														
CHEYENNE	75	69	73	67	69	63	57	46	54	61	66	71	75 NW	50
LANDER	65	77	62	72	56	61	57	56	70	70	75	72	77 SW	50
SHERIDAN	73	70	65	66	71	66	73	72	66	66	84	63	84 SW	50
PUERTO RICO														
SAN JUAN	46	43	41	35	45	40	62	80	149	43	40	44	149 NE	60

FIGURE 4-14. Continued

TWO BASIC TYPES OF WIND ROTATORS

There are two basic types of wind machines: the propeller type and the vertical rotator type. Propeller types are defined for the purposes of this book as those whose initial rotation is imparted to a horizontal shaft. Vertical rotator types are defined as those that impart their initial rotation to a vertical shaft. (See Figures 4-15 and 4-16.)

Very little proven effort has gone into vertical rotators while many propeller types have been used for centuries with quite a few brands available on the market. However, for the experimenter who is building his own this need not be a limitation. Our concern will be with obtaining the maximum electrical power from a given available wind in such a manner as to be economical, comply with local zoning restrictions, and not create an ugly sight.

Let's start from the beginning. First, you must have enough average wind to make a wind machine worthwhile. Second, you must know the maximum power that you want to obtain from that wind machine. Third, that maximum power ties you to a square-foot area of wind that you are going to use to get your power (based on certain efficiencies and wind speed, of course). Fourth, you must decide which type of wind machine you are going to build that will give you the maximum efficiency consistent with cost, local zoning, and appearance requirements.

WIND POWER EQUATIONS

Assume you have enough wind and you want to start on your wind machine. The maximum power available from the wind is determined by the equation:

$$P = K\tfrac{1}{2}\rho\ AV^3, \text{ where}$$

	P	= Power in watts
(RHO)	ρ	= density of air in slugs/cu.ft.
	A	= area (silhouette) of rotating components in sq. ft.
	V	= speed of the wind in ft./sec. (MPH × 1.47 = ft./sec.)
	K	= 1.356 (Conversion of ft.-lbs./sec. to watts)

FIGURE 4-15. PROPELLER TYPES

FIGURE 4-16. DARRIEUS USING SAVONIUS STARTING BUCKETS

The value of RHO (ρ) varies with altitude as below.

Condition: ICAO Standard Atmosphere

Altitude above sea level (ft)	ρ = **Density of Air** (**Slugs/cu. ft.**)
Sea Level	.002377
5,000	.002048
10,000	.001756
15,000	.001496

Apply this equation to sizing your system. We will make some assumptions for now on the amount of power you need and in Chapter 7, "Creating Electrical Storage," I will show you how to compute it. Let's say you want a peak power of 600 watts. This could be provided by a 1-HP, 12-volt, PM DC motor used as a generator. Normally, your maximum power output is planned for a 25-MPH ($\times$ 1.47 = 36.75 ft/sec.) wind. Now see what area your wind machine will have to be if it is 100% efficient in extracting the wind energy.

Power (of the wind) = $K\frac{1}{2}\rho\,AV^3$, our wind power equation

$A = \dfrac{2P}{K\rho\,V^3}$, rearranged algebraically to obtain area.

$A = \dfrac{2\,(600)}{(1.356)(.002377)(36.75)^3} = 7.5$ sq. ft.

Assume your wind machine is only 20% efficient, then you need $\dfrac{7.5}{.20}$ = 37.5 sq. ft. of wind machine area. That's 4 ft. $\times$ 9.375 ft. if rectangular and 6.9 ft. diameter if round. Six hundred watts sounds pretty good doesn't it? Well, don't be fooled—that only occurs at 25 MPH winds. In the equation, the wind velocity is cubed. That means that if the wind velocity doubles, the power increases 8-fold. Or in reverse if the wind is ½, the power is ⅛ of what it was. In our example of 600 watts at 25 MPH winds, the power is 75 watts at 12.5 MPH wind and 9.375 watts at 6.25 MPH wind.

Now work it all from a different angle. I'm going to put efficiency (E) into the wind power equation to get the power output from your wind generator.

Power (output) = $K\frac{1}{2}\rho AV^3E$; we will assume sea level (ρ = .0024); area = 154 sq. ft. (a 14-foot diameter); a wind speed of 10 MPH (14.7 ft./sec.) (remember ft./sec. = MPH $\times$ 1.47); and an efficiency of 50% (.50). I picked these values because they represent a feasible achievement on your part. We get:

$$P \text{ output} = K\tfrac{1}{2}\rho A V^3 E$$
$$= (1.356) \tfrac{1}{2} (.0024)(154)(14.7)^3(.50)$$
$$= 397 \text{ watts at 10 MPH}$$
$$\text{or } 137 \text{ watts at 7 MPH}$$
$$\text{or } 3,184 \text{ watts at 20 MPH}$$

Obviously, you can't get 3184 watts from a 600-watt generator, which means the extra energy will go into increasing the speed of the wind machine until it is either braked by some mechanism or it comes apart.

WINDMILL EFFICIENCIES

Requirements for designing vertical rotators and propeller rotators are very similar for area calculations with the exception that different efficiencies must be used. The Dutch four-bladed windmill has an efficiency of about 16%. The Savonius barrel-type vertical rotator has an efficiency of about 20%. The multibladed water-pump windmill has an efficiency of about 30%. The high-speed propeller type has an efficiency of about 42%. Some propeller wind machines have been built with an efficiency of more than 50%. I have probably missed a few types, but you get the idea. When you finish building your windmill, the efficiency can be computed from experimental results simply by recording the wind speed, volts and amps at the generator. Volts times amps equals watts of power on a DC system (Ohms Law). Compare this to the maximum power available from the wind and you get efficiency. This answer will vary with wind speed.

$$\frac{P \text{ (generator)}}{P \text{ (wind)}} \times 100 = \% \text{ efficiency}$$

Two types of windmills can be built: the propeller type or a vertical rotator. They are both about the same cost of $350 each. Plans for each are shown in later chapters. With these examples and this chapter information, you should be able to design different sized units to your needs.

WINDMILL COMPARISONS

Practical advantages and disadvantages of different types of windmills have been partially explained. We have discussed cost and efficiency. The remaining considerations are appearance and zoning restrictions. Appearance, zoning and height should be considered together.

Appearance is a combination of shape, color, height, and background. You have little control over shape except for the supporting structure which gives you some latitude. For example, you can pick a self-supporting structure or a guy-wired structure. The self-supporting structures are larger and more expensive but have no guy wires.

Color is something that you have complete control over and it can make a big difference. Depending on your location and desires: you may make your windmill gaudy, inconspicuous, pretty, bright, dull—you name it. It is not necessary to stick to one color either. For example, you might select all white (the laboratory look); or green, yellow and brown splotches (camouflaged); or red, white, and blue (bright and patriotic); or purple, lavender and white; or striped.

Height, especially in relation to surroundings, affects appearance. For example, a 10-foot tower on top of the rear of your house is less conspicuous than a 30-foot tower in the middle of your backyard. Your windmill will be about the same height above the ground to say nothing about the cost being less for the smaller tower! You will have to use your own judgement when it comes to the appearance that you like and which will not draw adverse criticism from close neighbors. You will have to keep other practical considerations in mind when deciding on appearance. For example, Mr. Landing (of the Chapter 11 design) does not recommend the propeller type on top of a house because the gear noise transmits into the house interior.

Zoning usually restricts where on your property you can build and how high. So before doing anything in the way of serious plans check on your zoning restrictions. This is controlled by city planning groups, usually, if you live inside city limits. If you live outside the city limits, you must consult county (parish or burrough) planning groups. If you don't know what name these groups may be under, ask your local building inspector—he is familiar with the codes he enforces and what groups lay them out. On my property I found the limits to building to be: nothing within 10 ft. of the property line except fences and a maximum structure height of 30 feet. Of course, TV antennas were exempt from the 30-foot restriction. So I asked if windmills were exempt also. They said they didn't know and would have to rule on the specific structure. Obviously, a Dutch windmill is a structure, but a propeller on top of a 40-foot ham antenna mast may be hard to rule against. A ham operator may erect such a structure with more antenna hanging at the top than there would be if it were a simple propeller and generator instead.

Whatever you do, don't run out and buy an expensive windmill and 90-foot tower and then check the restrictions. Some unfortunate soul in New York did just that! His windmill was still inside the garage the last I heard because the zoning restrictions made his particular installation illegal.

Height is another important factor to consider for your windmill. The advantage to getting it as high as possible is to get above the wind obstructions and to reach the higher velocity winds that are farther above the ground. There are equations to calculate wind changes above the ground due to height over smooth terrain, but these are meaningless where houses, trees, and hills exist. Therefore, a better rule is to aim for a height of 15 feet above all obstacles within 400 feet of the structure. That is easier said than done in some cases. If you live within 400 feet of a 400-foot hill obviously a 415-foot tower is not feasible. You may be able to find a spot on your property where the wind is maximum. Mount the windmill on top of a building or hill if you can. If you are not able to gain height that way, get as far away from obstacles as possible to install your tower. You may be able to find a highest wind location (perhaps caused by funneling among hills or trees) by observing the terrain over which you must select the location. Wind can be estimated by the following:

4-7   MPH: wind felt on face, leaves rustle
8-12 MPH: leaves in constant motion
13-18 MPH: dust and paper rises
19-24 MPH: small trees sway

On open terrain, watch the weed movement or place stakes in several locations and observe cloth strips tied to them. Try to do this when the wind is blowing from the direction that it usually does in your location (prevailing wind).

PRACTICAL ADVANTAGES AND DISADVANTAGES

For a summary of advantages and disadvantages by types of windmills for which design is shown in this book, see Figure 4-17. (In Figure 4-17, I have assumed certain sizes of equipment as recommended in the building chapters.)

TYPE	ADVANTAGES	DISADVANTAGES
VERTICAL ROTATOR	1. Modest cost 2. Lower profile 3. Easy generator attachment 4. Non-directional 5. Easy servicing 6. Lower starting wind speed	1. Unusual appearance 2. High weight 3. 20%-40% efficient
MULTIBLADED PROPELLER	1. 30%-50% efficient 2. Lighter in weight 3. Expected appearance 4. Modest cost	1. Higher profile 2. Directional 3. Difficult to service

FIGURE 4-17. COMPARISON OF WINDMILL TYPES

Which windmill should you select? I like the vertical rotator myself, but you may like the propeller type better. It's your choice and both work.

Now that you have selected a windmill type to build, see Chapter 10 for a discussion on generator selection; alternators have been ruled out for the reasons given there. Permanent magnet field DC motors are recommended as generators. Others may be substituted if you have them available. What you want in a generator, under ideal conditions, is maximum power at low RPM, weatherproof, ball or roller bearing on one or both ends, brushless and rugged. In reality, you may settle for a weather-resistant type that has none of the other features mentioned above, but is only $100 compared to $400 for one with all the desirable features mentioned above.

Voltage regulation is not absolutely necessary if you have low power output from your windmill. However, if you have significant power levels in high winds (over 500 watts), you need voltage regulation to prevent system or battery damage. (The selection of voltage regulation components will be discussed in Chapter 9.)

Connection of wind systems to batteries is accomplished by paralleling or series connecting batteries to obtain the voltage and amp-hour storage capability desired. An example is shown in Figure 4-18 for a 12-volt generator and 12-volt battery voltages. Exact calculations will be discussed in Chapter 7 on creating electrical storage.)

Liability insurance may cross your mind as a possible expense. I checked with one insurance company and their representative said that windmills are covered the same as other hazards such as wells and ditches, probably at no additional cost to your present policy. Another insurance company said it was not necessary to notify the company of the windmill addition as it would be treated the same as a TV antenna, unless some specific exception was made in the policy. To be sure where you stand on liability insurance, contact your insurance company. Expect no problem nor additional cost for coverage.

You were told that wind was the cheapest source of electric power. You received an education on windmill types, you were shown how wind velocity drastically effects your capability to obtain wind power. Costs, efficiencies, appearance, zoning, and height were discussed in realistic terms. The decision to build a windmill rests on wind availability, cost considerations, and all those advantages and disadvantages you must weigh. If you should choose to build a windmill, building details are given in Chapters 10 and 11.

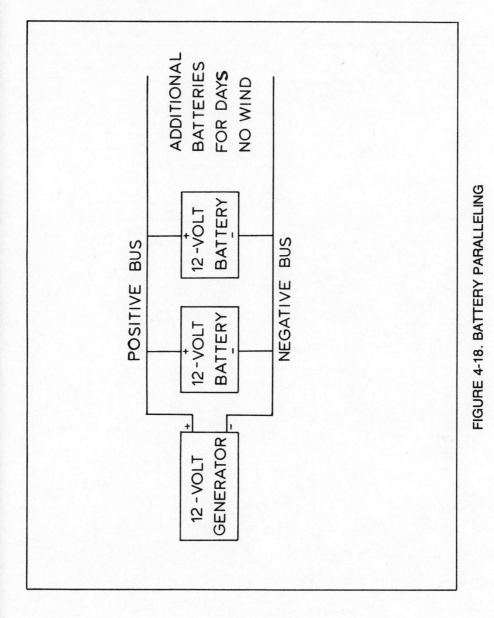

FIGURE 4-18. BATTERY PARALLELING

5.

Using the Sun Directly

The sun is the primary source of free energy on the earth. An understanding of the sun and its energy output are necessary to properly use that energy. In this chapter, use of the sun's energy through photovoltaic methods to obtain electrical energy will be discussed in detail. Mirror-target methods of using the sun's energy will be discussed here and in Chapter 14, but tested designs for use of mirror-target systems are beyond the intent of this book.

HOW THE SUN CREATES ENERGY

The sun is a gaseous sphere, 864,000 miles in diameter and is an average distance of 93 million miles from the earth. The sun rotates on its axis about once a month. The approximate surface temperature of the sun is 5,762°K. The density in the center is about 100 times that of water. The sun is a continuous fusion reactor using its own gases as a containing vessel. The primary fusion function is believed to be that of four hydrogen protons forming one helium nucleus with the mass lost in the process given off as energy between 8 and 40 million °K in the central interior. It is esimated that 90% of the sun's energy is generated in the central 23% of the diameter where 40% of the mass exists.

HOW MUCH ENERGY REACHES THE EARTH

Viewing the sun from the average earth distance of 93 million miles, it subtends an arc of 32 minutes (0.53 degrees). Our distance from the sun varies ± 1.7% during the year. This causes the energy reaching the earth to vary ± 3% during the year. The peak energy occurs January 1st and the lowest on July 1st. According to NASA's spectral irradiance data, the energy versus wavelength curve for energy received outside the earth's atmosphere at the sun's average distance from the earth is shown in Figure 5-1.

The total of the sun's energy over the full frequency spectrum reaching the earth's outer atmosphere at the average earth-sun distance is called the ''solar constant.'' Based on Figure 5-1, the solar constant is 1353 watts per square meter (428 BTU/ft²hr or 125.7 watts/ft²). The actual energy varies from about 1398 watts/m² on January 1st to about 1310 watts/m² on July 1st in almost a perfect sine wave change.

Now that we know what sun energy reaches the earth's outer atmosphere, the real question is how much of it reaches the earth's surface. This is a variable that depends on water vapor, dust, miscellaneous gases present, latitude, time of day, time of year, and weather conditions. Ozone in the upper atmosphere and water vapor in the lower atmosphere filter almost completely all ultraviolet and infrared energy outside of a band pass of .3 to 2.3 micrometers wavelength.

All daily data for the sun's energy at a specific location are given in solar time. This is the time when 12 noon is said to occur just as the sun passes directly overhead at your longitude. In each time zone, there is a longitude where local standard time equals solar time. For the continental U.S. these are 75° W for Eastern, 90° W for Central, 105° W for Mountain, and 120° W for Pacific time. For all other longitudes within a time zone, solar time varies from local standard time by less than one hour. The difference can be computed, but for the purposes of this book it is sufficient to note that solar time is within one hour of local *standard* time.

It is not practical to predict weather for forecasting sun energy. Therefore, the approach I will use will be to give examples of the sun reaching the earth on clear days and to give mean total hours of annual sunshine in the United States based on sunshine records from the Weather Bureau over a 29-year period.

As you can see from Figures 5-2 and 5-3 a normal (perpendicular) surface to the sun's rays does increase energy received over a horizontal surface. The locations of these figures are different and

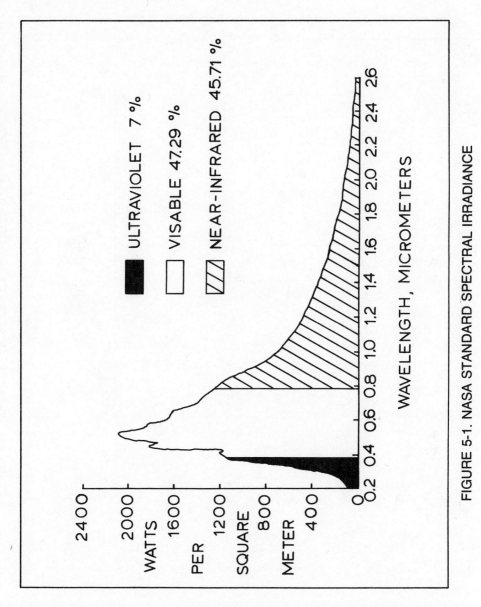

FIGURE 5-1. NASA STANDARD SPECTRAL IRRADIANCE

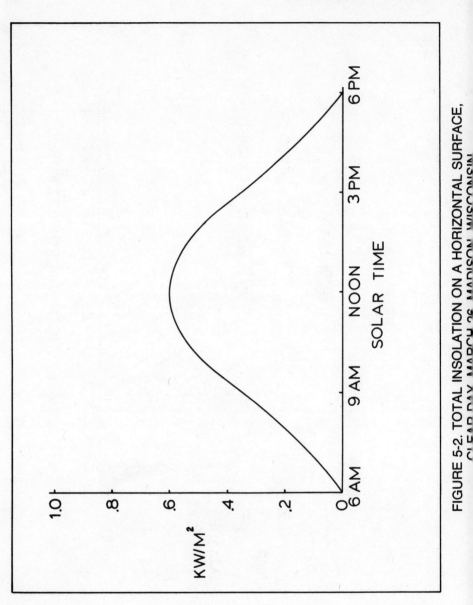

FIGURE 5-2. TOTAL INSOLATION ON A HORIZONTAL SURFACE, CLEAR DAY, MARCH 26, MADISON, WISCONSIN

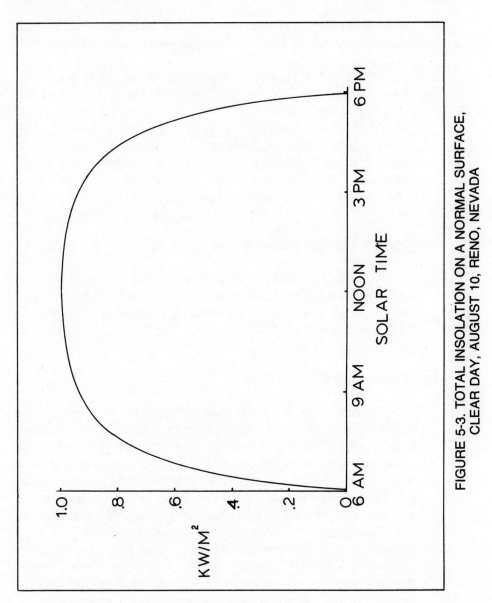

FIGURE 5-3. TOTAL INSOLATION ON A NORMAL SURFACE, CLEAR DAY, AUGUST 10, RENO, NEVADA

therefore not directly comparable, but the principle shows. The term *insolation* means the *sun's energy* and is not to be confused with insulation.

In Figure 5-4, curves of annual hours of sun are shown for the United States. This will give you a good idea of the annual insolation available in your location.

In the United States, a *peak* insolation power at noon of 1,000 watts per square meter can be assumed as a standard. Even though it may not be reached in all areas at all times, it makes a nice design reference. It can be restated as 100 milliwatts per square centimeter (100 mW/cm^2) or 645 milliwatts per square inch.

METHODS OF ENERGY CONVERSION

Now that you have an idea of what energy is available, you should know that it can be converted to electricity by only two ways: photovoltaic conversion or heat energy conversion to motion to drive a generator.

HEAT-TO-MOTION CONVERSION

The heat-to-motion conversion method is the most difficult but shows the most promise for solar-electric power plants of the future. In this method, the sun is received by mirrors and reflected onto a solar collector. Many flat mirrors reflecting their energy onto a single solar collector can be said to focus the energy on the collector. The collector is so designed that liquid or gas flowing through the collector is boiled or super-heated. This heated gas is then directed on turbines (such as a steam turbine) which drives a large generator. Such power plants should have an efficiency of about 25% which is normal for solar-electric power plants. Such a system can have many variations in design but the basic principle requires the mirrors to track so that their reflected energy always falls on the desired target receiver. This means a flat mirror always points (is normal) to a position half way between the sun and the receiver. Under these conditions, angular tracking speed of the mirrors is always one half the angular speed the sun moves across the sky. A

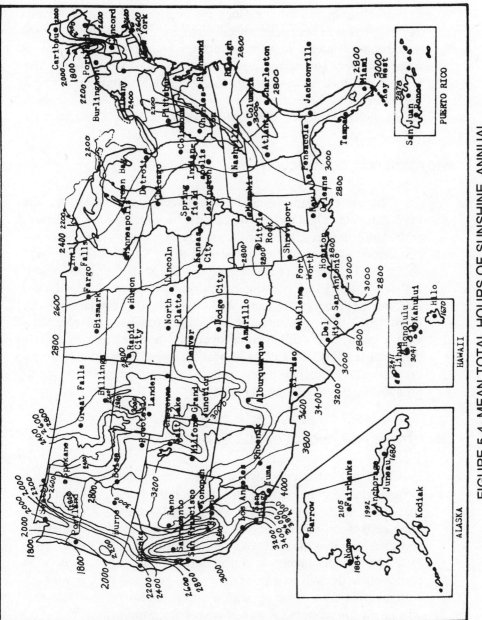

FIGURE 5-4. MEAN TOTAL HOURS OF SUNSHINE, ANNUAL
(Based on records between 1931 and 1960; U.S. Weather Bureau)

computer is required for open-loop tracking since it varies for every day of the year in both azimuth and elevation. A closed-loop control may be used where sensors correct the mirror reflections to the target receiver. The mirrors could be focused on a steam boiler. The boiler could then be used to run a steam engine or turbine driving a generator. The generator would be sized at 250 watts for each square meter of mirrors. (General design is provided in Chapter 14.)

PHOTOVOLTAIC CONVERSION

The use of photovoltaic devices is technically very easy. The most common cell is the silicon cell used for many years on spacecraft. Unfortunately, these are relatively expensive. The ones I experimented with were sold by Calectro (Cat. No. J4-803) and were $6.50 for 12 square centimeters (about ¾" × 2¼" or 1.86 square inches). That's $3.50 per square inch! These were only 7% efficient in converting energy to electricity. Although silicon solar cells can reach 12 to 14% efficiency, this must happen with selected cells with a good heat sink. Normal operating temperatures of silicon cells can reach 60°C (140°F) in peak sunlight. These temperatures decrease the efficiency of the cells. Therefore, it is important to provide the best heat sink that you can.

THE FUTURE FOR PHOTOVOLTAIC DEVICES

Although silicon photovoltaic cells are the most common cells available in large quantities so far, the big disadvantage is cost. The cost of the silicon cells I bought was $78 per watt of peak output and these cells were unmounted and unprotected! From the right source and in reasonable quantity you can obtain them for $30 per watt. Don't be too disappointed. Since virtually the only drawback to photovoltaic devices as a power source is cost, industry is working extremely hard to come up with a cheap photovoltaic device that will create an economical power source for homes.

For example, Varian Associates in California found an efficiency breakthrough by developing a gallium arsenide photovoltaic cell with 20% efficiency that is capable of operating at temperatures where focused energy could be used. They said a cell ⅓ of an inch in diameter could produce 10 watts. This author computes that a focusing surface of 77 sq. in. must be used to focus on the ⅓ inch diameter cell and the unit must track the sun. This may also require a cooling system, but Varian didn't say. Previous gallium arsenide cells were known to have only 10% efficiency.

Solarex of Rockville, Maryland, developed a new kind of silicon solar cell that could be the price breakthrough that everyone is looking for. The new cell is made of polycrystalline silicon (a less pure silicon). Solarex produced a cell that is 10% efficient; very impressive for polycrystalline silicon.

In the common silicon cells now in use, the cell is composed of two flat crystals of silicon pressed together; one is doped with impurities to produce positive charges and the other doped with different impurities to produce negative charges. These opposite charges (created when light strikes) cause a current to flow. This is the cell output. Should the crystals be impure silicon, the current would be stopped by irregularities in the crystal. Solarex has claimed a process by which the impure crystal material used does not stop the current flow but rather only decreases the flow from 14% efficiency down to 10% efficiency. Although it may be difficult to mass-produce it at 10% efficiency, I personally would settle for anything close to that with a lower price than the current cells available. The polycrystalline silicon *material* can be obtained for $1/10$ of the cost of the pure silicon crystals now used. But since the crystals only represent 15% of the cost of solar cells (the remainder is labor), this is only a 13% price reduction. Couple this with automation in manufacturing and price could really come down. To be economical, automation will require a larger market than we now have for the cells. Solarex states that a *volume* buyer can now obtain silicon solar cells from them at $17 per watt and they expect the price to go to $2 per watt by 1981 *assuming a triple increase in annual sales volume each year*.

These are just examples of where the market is going. With the price of photovoltaic cells going for $17 to $75 per watt (depending on quantity), it is currently not economial to power your whole house with this source. It is estimated that about 1,000 square feet of silicon cells (providing 5,000 to 6,000 watts) would be sufficient to power an average home entirely. Such a system would charge batteries, and inverters would convert the battery power to 120 volts AC for standard house use. However, at present the cost of the silicon cells would be $100,000. This price must drop to under $10,000 to make the cost feasible. The price per watt must drop to $1.70 to make home power economical on a large scale. This could happen by 1982. (However, due to inflation the right price in 1982 may read $3.00 per watt.)

While prices are coming down, I recommend a trickle charge capability. My 1-watt system cost $75.00 and will put out about 144 watt hours per year. That is only one cent of electricity per year. At that rate it would pay for itself in 7,500 years! See my point? But, don't be discouraged. Photovoltaic devices are still my favorite forecast for homegrown power in the 1980's.

HOW TO USE PHOTOVOLTAIC POWER

Should you decide to experiment with photovoltaic devices see Figures 5-5 and 5-6. A silicon cell such as the one in Figure 5-5 typically puts out 0.45 volts and 50 milliamps in bright sunlight at noon. Typical current-voltage curves are shown in Figure 5-6 at various light levels for the cell shown in Figure 5-5. If the load resistance is very low, the cell acts as if it is shorted at the output and the current varies in proportion to the amount of light falling on it. If the load resistance is very high, the cell acts as if it is open-circuited and the voltage rises very rapidly to maximum voltage. Current at that voltage is then limited by the amount of sunlight and the load resistance. This characteristic is ideal for charging batteries. Using 2 cm × 2 cm silicon cells to charge a 12-volt battery (figuring .3 volt each) will require 40 cells connected in series to create a solid charge level. However, if you are short on quantity, 27 cells in series will begin to charge a battery that is at 12 volts. Higher current will result with 40 cells. When multiples of 40 cells are available, parallel them for higher current output. When connecting cells in series, connect the red wire on one cell to the black wire on the adjacent cell as in Figure 5-7. Put a diode in series with the battery's positive terminal. This will prevent reverse current flow (a small battery drain) when the cells are not receiving sufficient light to charge the battery. When connecting cells in series,

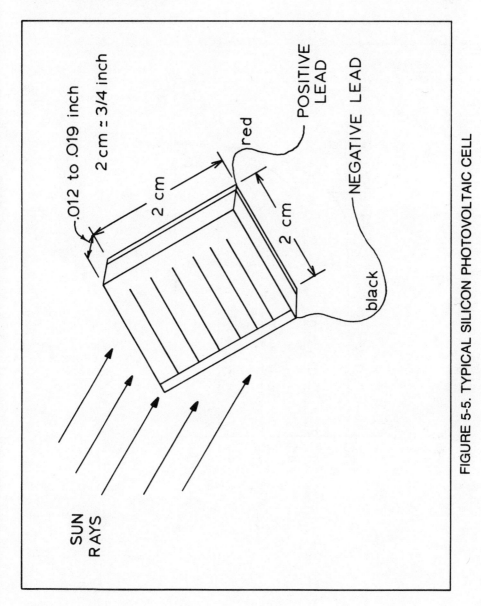

FIGURE 5-5. TYPICAL SILICON PHOTOVOLTAIC CELL

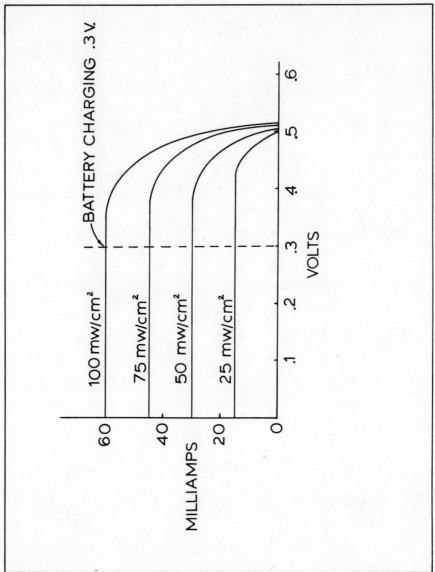

FIGURE 5-6. TYPICAL CURRENT-VOLTAGE CHARACTERISTICS

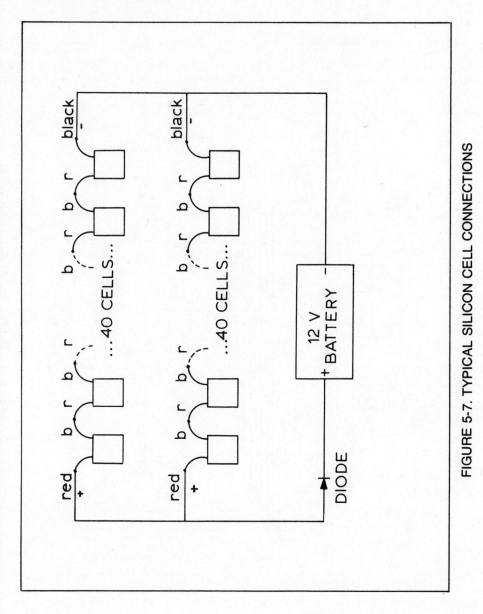

FIGURE 5-7. TYPICAL SILICON CELL CONNECTIONS

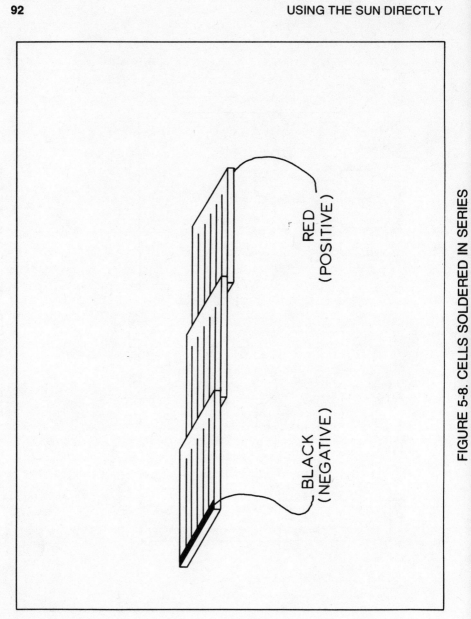

RED
(POSITIVE)

BLACK
(NEGATIVE)

FIGURE 5-8. CELLS SOLDERED IN SERIES

connect the wires as just mentioned. Manufacturers sometimes solder cells together as shown in Figure 5-8. It is very easy to ruin a cell attempting to solder it, therefore directly soldering cells by individuals without special equipment is not recommended.

Mounting of cells should be to a heat sink (usually an aluminum plate) with heat conductive but electrically insulated compounds or glues. This will reduce operating temperature and make the cells more efficient. If you have a free water source you could water cool the heat sink.

The negative sides of the cells usually face the sun and have anti-reflection coatings. These coatings should be protected from dust, bird droppings, ice, etc. by using a clear plastic or glass cover. Accumulated dust on the cover will reduce the output about 10% and should be rinsed occasionally. Do not seal the cells completely—leave an air vent (a very small hole) where rain can't get in to prevent pressure build-up inside due to temperature changes. Current output for light levels can be increased by reflectors.

The cells should face south for fixed applications north of the equator. The angle off the ground should be equal to your latitude for year around average or can be changed monthly to face the sun at noon for more efficiency.

Direct use of the sun is expensive, but not impossible. Keep experimenting, cheaper equipment is on the way.

I told you how the sun creates energy and how much sun reaches the outer atmosphere. Then you learned the variables involved in getting energy to the earth's surface. Finally, you were told that only two methods were available to obtain electricity from the sun's rays: photovoltaic conversion and heat-to-motion conversion. Of these methods, photovoltaic conversion is simple and heat-to-motion conversion very complicated. Unfortunately, both methods are extremely expensive. I have forecasted less expensive photovoltaic devices for the early 1980's.

I recommend experimentation in photovoltaic devices so that you will be ready for these units when they become available at reasonable prices. A good project is a small battery charger. Some specific guidelines on what to expect and how to select the correct number of cells is in order. When looking at specifications on photovoltaic devices (silicon or otherwise), you will see three basic ratings: open circuit voltage (V_{oc}), short circuit current (I_{sc}), and per cent efficiency. Usually, these are all measured at the 1000 watts per square meter input level (peak power). You can measure all of these outside at noon (solar time) yourself.

Remember that open circuit voltage and short circuit current don't give you operating parameters. However, there is a simple way to compute operating parameters. It is based on the fact that until you exceed battery voltage, no current is going to flow to the batteries from the photovoltaic cells. Let's take an example that illustrates this point. I am using 2 cm × 2 cm silicon photovoltaic cells; 33 of them. Each one is rated at 0.45 V_{oc} and 50 ma I_{sc}. I wish to charge a 12-volt battery. Therefore, I connnect the silicon cells in series. This gives a total V_{oc} of 33 × .45 = 14.85 VDC and I_{sc} remains at 50 ma when connecting in series. What current will I get when I connect these across the battery? Let's say the existing battery voltage is 11.5 volts and that your series diode that you will use (see Figure 5-8) has a dropping voltage of 0.6 volts. No current will flow until your photovoltaic group reaches 11.5 + 0.6 = 12.1 volts. Assuming a peak sun input, the current that flows to charge the battery is:

$$I_{charge} = \frac{(V_{oc} - V_B)\ I_{sc}}{V_{oc}}$$

where V_{oc} = total open circuit voltage of the photovoltaic cell group

V_B = battery voltage (including series diode)

I_{sc} = short circuit current of photovoltaic cell group

In the example case:

$$I_{charge} = \frac{(14.85 - 12.1)\ (50\ ma)}{14.85} = 9.26\ ma$$

What we did was subtract the opposing battery voltages from the available voltage to get a ratio of the available current. In practice, I found this to be the exact charging current assuming the battery doesn't change voltage. If the battery voltage goes up due to charging current, you must substitute the new battery voltage in the equation. This shows that current levels fall off as the battery approaches its full charge.

In the example case we had only 2.75 volts available above battery and diode voltages. With .6 volt being lost in the diode, it is tempting to leave the diode out of the circuit. If you do, you will have a small reverse leakage current of 20 to 50 microamps per series string of photovoltaic cells during the night and bad weather periods. If this is detrimental to your efforts, or to other charging circuits that are charging the battery, you will need a diode. To get around the 0.6-volt voltage drop on silicon diodes, you can use a germanium diode. However if you are putting out very much current from your photovoltaic group, you will have trouble finding a germanium diode rated at a high enough current. Instead use a *germanium* PNP power transistor and short the base to the collector. The emitter will connect to the positive output of the photovoltaic cells and the collector will connect to the battery positive terminal. Maximum collector current will be the diode rating.

To select the correct quantity of photovoltaic cells in a series string, divide the normal battery voltage (battery to be charged) by the recommended photovoltaic cell voltage for charging batteries. If a battery charge voltage is not given for your photovoltaic cells, use 75% of the open circuit voltage. This will give you the number of cells in a series string to charge batteries at about the maximum power output. If you desire more current, add parallel strings of cells until you reach the desired level.

For example, using a photovoltaic cell rated at 0.45 volts open circuit voltage and 0.3 volts recommended battery charge level with a 50 ma short circuit current rating, how many cells and in what configuration would you need for charging a 12-volt battery at a 100 ma rate?

No. of cells in series strong $= V_{Battery} \div V_{BC}$ of cell
$$= 12.0 \div 0.3 = 40 \text{ cells in series string}$$
where V_{BC} is the battery charging voltage of a cell

$$I_{charge} = \frac{[(V_{oc} - V_B)] I_{sc}}{V_{oc}} = \frac{[(.45 \times 40) - 12.0] (50 \text{ ma})}{(.45 \times 40)}$$

$$= 16.67 \text{ ma per series string}$$

No. of series strings $= \dfrac{I_{desired}}{I_{series\ string}} = \dfrac{100 \text{ ma}}{16.67 \text{ ma}} = 6$

Total number of cells $=$ (cells in string)(No. of series strings)
$$= 40 \times 6 = 240 \text{ cells}$$

The efficiency rating of the photovoltaic cell is usually given. If it is below 10%, you probably have a low quality cell. If it is not given or if you wish to check it, an approximate efficiency can be obtained experimentally by taking measurements at solar noon on a clear sunny day. Under these conditions, you may assume that 100 mW/cm² is reaching your test cells when they are perpendicular (normal) to the sun. Load your cells with a variable resistor until they are at battery charging voltage or a maximum (current × voltage) wattage output. Record the current (I) and voltage (V). Compute efficiency:

$$\% \text{ Efficiency} = \frac{(I \text{ in ma})(V \text{ in volts}) (100)}{(\text{area in cm}^2) (100 \text{ mW/cm}^2)}$$

This is only an approximation. More accurate efficiency may be obtained by measuring the light input in a laboratory. The efficiency tells you how much of the sun's energy striking your cells is changed to electricity.

6.

Unlimited Free Energy Sources

In this chapter you will learn how to use moving water, muscle power, potential energy, pressure, and other possible energy sources to make electricity.

HOW TO USE MOVING WATER

If you are considering using moving water to make electricity, you should know the federal laws concerning water. No laws apply to the use of streams and nonnavigable rivers on private property. This is confirmed by the U.S. Department of Interior, Bureau of Reclamation, Water and Land Use Division, Water Operations Section. The Federal Power Commission advises if a water wheel is installed in navigable waters it requires a license from that Commission and if the water wheel is located in a National Forest or other federal lands, it requires a license from the Department of Interior's Bureau of Land Management. The Bureau of Reclamation will be interested in your operation if you try to sell your electricity. Private use requires no licensing or permissions on nonnavigable waters on private land. It is unlikely that any local ordinances will apply either, but this is left to the builder to check with his local building inspector.

Moving water can be used to rotate either a vertical shaft or a horizontal shaft. Since the vertical shaft requires a small, fast-moving body of water, it will not be discussed here. The horizontal shaft water wheel will operate in almost any moving water situation. (See Figure 6-1.)

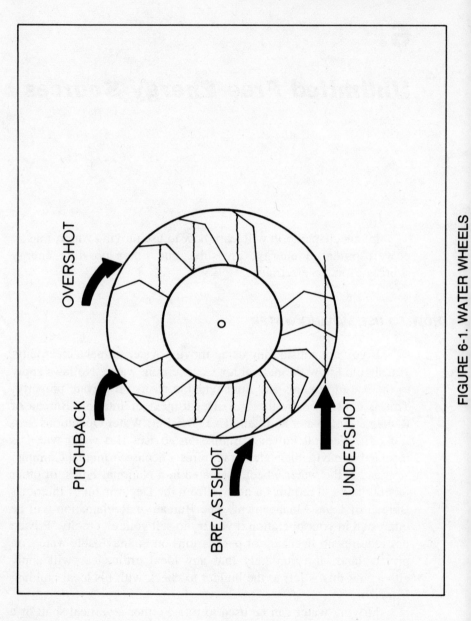

FIGURE 6-1. WATER WHEELS

Depending on at which point the water enters it, the wheel is called overshot, pitchback, breastshot, or undershot. The overshot is the most efficient, but requires getting the water to about 6 inches above the wheel. The water should enter the buckets just after the wheel passes the top position. The buckets should only fill ¼ to ⅓ to reach optimum efficiency. The amount of water entering the buckets can be controlled by a gate on the penstock (sluice). Overshot wheels need a head (difference in height of entering water to exiting water) of 4 to 30 feet, and a water flow of 60 to 1800 cu. ft. per minute because the power available in a stream is dependent on head and flow. An overshot water wheel is 60% to 80% efficient and turns at 2 to 12 RPM's. By the time you gear-up the RPM's for a generator, about 1 to 100, more power is lost and the efficiency falls.

Water turbines can be used if heads of 10 to 50 ft. or more are available. The two water turbine types are impulse and reaction. Impulse types use a nozzle to shoot water into buckets on a wheel. Reaction types bring water through fixed vanes onto moveable vanes forcing the moveable vanes to rotate a shaft. These turbines are considered to be high speed in comparison to water wheels. Because the turbines must be balanced and special metals used to reduce erosion, I recommend that they be purchased already built from a vendor such as Small Hydroelectric Systems in Custer, Washington.

CALCULATING WATER POWER

If you elect to build a water wheel, I recommend the overshot type. In computing available power from your water, the equation is:

$$HP = \frac{W \, Q \, H}{33,000} = \frac{(62.4)(\text{water flow})(\text{head})}{33,000}$$

where HP = horsepower

W = weight of water in pounds per cubic foot,

at 32°F, W = 62.42 lbs/cu. ft.

at 40°F, W = 62.42 lbs/cu. ft.

at 50°F, W = 62.41 lbs/cu. ft.

at 60°F, W = 62.37 lbs/cu. ft.
at 70°F, W = 62.31 lbs/cu. ft.
at 80°F, W = 62.23 lbs/cu. ft.
(for all practical purposes, W = 62.4 lbs/cu. ft. may be used)
Q = flow of water in cubic feet per minute
(more on calculating this follows)
H = head in feet (difference between height of
entering water and exiting water
to water wheel).
1/33,000 = horsepower conversion unit.

Calculating the flow of water available can be accomplished by several methods. If the stream flow is small it can be funneled entirely into a bucket of known capacity (1 gallon = 0.13 cu. ft.), then you can time the filling of the bucket to obtain cubic feet per minute.

The method I use for approximation of available water flow is to measure the cross sectional area of a stream at five evenly spaced points along the stream. Then I time a floating object along the portion that I have measured the cross sections in. For example, in Figure 6-2 where the lines of x's are shown, I made width measurements in inches. Along each width I measured depths in inches at equally spaced intervals. Each measured cross sectional area is equal to:

Position A:

$$36''W \left(\frac{4''D + 6''D + 7''D + 3''D}{4} \right) = 36''W \times 5''D_{Avg} = 180 \text{ sq. in.}$$

Position B:

$$30''W \left(\frac{6''D + 8''D + 7''D}{3} \right) = 30''W \times 7''D_{Avg} = 210 \text{ sq. in.}$$

Position C:

$$24''W \left(\frac{7''D + 9''D + 8''D}{3} \right) = 24''W \times 8''D_{Avg} = 192 \text{ sq. in.}$$

Position D:

$$32''W \left(\frac{3''D + 4''D + 6''D + 5''D}{4} \right) = 32''W \times 4\frac{1}{2}''D_{Avg} = 144 \text{ sq. in.}$$

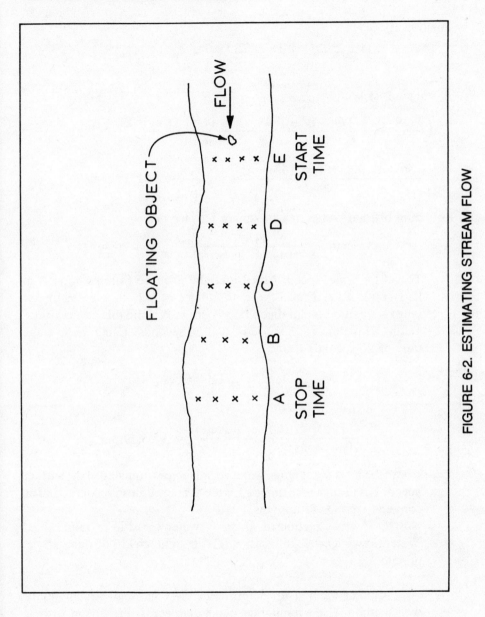

FIGURE 6-2. ESTIMATING STREAM FLOW

Position E:

$$30''W \left(\frac{4''D + 5''D + 8''D + 7''D}{4} \right) = 30''W \times 6''D_{Avg} = 180 \text{ sq. in.}$$

Average cross section   =

$$\left(\frac{POS. \; A + POS. \; B + POS. \; C + POS. \; D + POS. \; E}{5} \right) =$$

$$\frac{180 + 210 + 192 + 144 + 180}{5} = \frac{906}{5} = 181.2 \text{ sq. in}$$

Rounding off and converting to square feet we get:

$$\text{Average Cross Section} = \frac{181 \text{ sq. in.}}{144 \text{ sq. in./sq. ft.}} = 1.26 \text{ sq. ft.}$$

To get the water speed, I measured the distance from position A to E. I got 37 feet. Then I threw in a piece of bark (about 1½ inches long) at Position E and timed it to Position A. I did this several times getting 18 seconds, 22 seconds, and 20 seconds. Using the average time of 20 seconds I got:

$$\text{Water Speed} = \frac{5}{6} \frac{L(60)}{T} =$$

$$\frac{5}{6} \frac{(37 \text{ ft.})}{20 \text{ seconds}} \frac{(60 \text{ sec/minute})}{} = 92.5 \text{ ft/min.}$$

where the $^5/_6$ is the average water speed factor compared to the surface speed, L is the distance timed, T is the average time in seconds, and 60 converts seconds to minutes.

Now Q = cross section in sq. ft. × water speed in ft./min.

Water flow = Q = 1.26 sq. ft. × 92.5 ft./min. = 116.55 cubic ft. per minute.

Another method for determining flow in small streams is the weir method. Use a temporary plank dam across the stream with an accurately cut, horizontal, rectangular notch in its top edge (midstream). The amount of water in cubic feet per minute is related to the depth of water flowing over the bottom edge of the weir notch bottom edge (as measured a few feet upstream) and the width of the notch. For each inch of width the following flow rates apply:

Depth over Weir, Inches	Cu. Ft./Min. for 1 Inch Width
1	0.40
2	1.13
3	2.07
4	3.20
5	4.47
6	5.87
7	7.40
8	9.05
9	10.80
10	12.64

Example: 3″ deep water over weir, 10″ wide weir = 20.7 cu. ft/min.

Now let's carry the example through the horsepower equation.

$$HP = \frac{W \, Q \, H}{33,000} = \frac{(62.4)(116.55)(\text{one foot head})}{33,000} = 0.223 \, HP$$

This shows that almost ¼ horsepower is available for each foot of head. Slightly over 4 feet of head will provide 1 horsepower. However, the efficiency of the water wheel in converting to electricity must be considered. You can expect to claim 50% of the available power in an overshot wheel. The 50% lost includes water spillage, friction, generator losses. Assume that one horsepower was available at 50% efficiency. That's ½ horsepower of electricity and in watts:

$$\text{Watts} = HP \times 746 = ½ \times 746 = 373 \text{ watts}$$

Keeping in mind that for direct current (DC), volts times amps equals watts, at 12 volts you could obtain 38 amps continuously as long as the stream holds out. That is a lot of power! (In Chapter 12, an example of how to build a water wheel is provided.) If you have a stream, take advantage of it.

HOW TO USE MUSCLE POWER

Your own muscle power is another good way to obtain electricity, especially if you enjoy exercising to keep trim. Leg muscles are the most powerful group of muscles in the human body. Because this is so, a bicycle exerciser is the most useful means of producing electricity. It has been said that a human being is worth about ⅓ horsepower. (Here, the reference is to peak power. A much lower level should be used for working out equations with a bicycle exerciser. See Chapter 13 for further explanation.) Power output capacity fluctuates greatly between men and women, and between large and small people.

A bicycle exerciser is fairly simple to make. Arrange the bicycle to drive a generator with its rear wheel; the gear ratios necessary will depend on the generator's design RPM. Do not try to drive a one horsepower load with your bicycle. This will make you exhausted after a few minutes of getting some power and you won't want to use it again. Instead, design your exerciser for a very light load and plan on 15 to 20-minute exercise periods. This will create a reasonable amount of electrical charging to a battery and still not turn you off on the idea. Remember the more you use it, the better shape you will be in. An ammeter on the output to see what you are accomplishing is a way of holding a steady exercise rate. (Chapter 13 tells how to build a bicycle exerciser).

HOW TO USE POTENTIAL ENERGY

Potential energy is more difficult than the other sources of power to use but possible. Potential energy is defined as weight at a height. For example, if a 200-pound man descends an 8-foot high flight of stairs, at the top the potential energy is $200 \times 8 = 1600$ pound-feet. At the bottom the potential energy is zero. Why not put the 1600 pound-feet to work making electricity? Let's put the 200-pound man at the top of the stairs, take away the stairs, give him an elevator with nothing to hold it back when the brake is released. Obviously, the elevator will fall when released. So let's gear a generator to it. The faster the elevator tries to go, the more electricity is put out at the generator. Proper gear ratio to the generator will give a steady, slow speed to the next level down.

We have just created a one-way elevator. To make it practical, it must raise a second car while you are descending in the first, and you must use stairs to climb back up. However, a car will always be at the top to take you down. The economics of obtaining electricity in this fashion are questionable, but you can see that rather than having potential energy going to waste, it can be harnessed.

USING STORED PRESSURE

Pressure, if available for free such as in an artesian well, can also be harnessed to produce electricity. For water pressure, a water pump could be geared to a generator and the water pressure passed through the pump to waste or to be used as a pressure drop in a system with excessive water pressure. Depending on the water flow, a given pressure drop occurs across a water turbine, which is a pump in reverse. This pressure drop will increase with the size of the generator you gear it to. I'm not going to tell you exactly how to build one of these, but I do intend to make you think of all the energy around you that is going to waste. Just maybe, you'll run across a super energy source that is now going to waste in your vicinity that can be harnessed to your electrical benefit!

IDENTIFYING OTHER FREE ENERGY SOURCES

Look back a minute at the definition of free energy sources— anything that *heats* or *moves* can be one. I looked around my home and tried to come up with a list of free energy sources. Other than the ones I have already mentioned, I ran across the heat being dumped outside from the clothes dryer. Some problems exist in using this heat: it is moist; it cannot be back pressured; it contains lint. I haven't figured out how to use it *yet*, but there it goes—*to waste*. Electrically, I'm stymied, but an appropriate heat exchanger may allow me to heat part of the house with it in the winter. Think about the free energy that is going to waste! Look around. The only limitation is your ability to identify and use it.

This completes the energy source discussions. We will now proceed with how to use energy effectively and how to build your own energy systems.

7.

Creating Electrical Storage

Batteries are the only practical way of storing electricity. Other ways are possible but are not yet perfected. For example, a flywheel can be used to store energy (driven by a motor to store and driving a generator to retrieve power). However, practical efficiency is yet to be marketed on flywheel storage. Exchange media can be used to replace direct storage as utility companies do (buying power from another utility company when needed). Some water storage is used to store electricity such as at Mt. Elbert Pumped-Storage Powerplant in Colorado. When excess power is available, it is used to pump water into a reservoir from a lake. The reservoir is 300 to 500 feet above the lake. When peak demands hit, the reservoir is emptied through the power plant, driving the motor as a generator. However, for private use, batteries are the best solution to electrical storage presently.

COST VS LIFE OF BATTERIES

Cost, life, and ampere-hour rating of batteries are linked closely. Figure 7-1, a chart of automobile batteries of the lead-acid type, will illustrate this point. As you can see, when the cost goes up, the warranty life and ampere-hour rating increase. The 6-volt battery may throw you a curve unless you realize that you need two 6-volt batteries to compare to a 12-volt battery. Furthermore, when adding batteries in series to increase voltage, the batteries used must

RATED VOLTAGE	AMP-HOURS @ 20 HR. RATE	WARRANTY LIFE	APPROXIMATE COST - $
12 V	100	Lifetime of car	46
12 V	95	6 yr.	42
12 V	85	5 yr.	36
12 V	75	4 yr.	30
12 V	50	3 yr.	26
12 V	32	2 yr.	22
6 V	85	2 yr.	20
Series Two 6 V (as above)	85	2 yr.	40

FIGURE 7-1. AUTO BATTERIES

be the same amp-hour rating. The result is a higher voltage with no increase in amp-hour rating. (See the example of two 6-volt batteries in Figure 7-1.)

WHAT BATTERY TYPES ARE AVAILABLE

Auto batteries are not the only source of batteries. They lack two things: amp-hour ratings much over 100 and design for deep discharge. Truck batteries can be obtained at 200 amp-hour ratings, but their capacity for deep discharge is about the same as auto batteries. (Both are designed for alternators or generators to maintain them near full charge except on rare occasions.) In the event deep discharge is expected frequently in your system, electric golf cart batteries are available and these are designed for deep discharge. They can be purchased from golf cart dealers or battery outlets. Most are 6-volt batteries. A typical unit is 6-volt, 200 amp-hours, 2-yr. warranty, at approximately $68. Keep in mind that it takes two of these for a 12-volt system at 200 amp-hours. 12-volt, deep discharge recreational vehicle batteries also are available: 105 amp-hours, 2 year warranty, $75 (example Gould PB-105).

Commercial uninterruptable power systems use batteries that have extremely long life. Lead-calcium and nickel-cadmium batteries can be obtained with 20-year warranty life. The initial cost is high. For example, a 120-volt, 400 amp-hour battery group may cost over $5,000. That's $250 per year prorated for a 48 kilowatt-hour storage. If such battery systems interest you, get several price quotes before you buy.

HOW TO CALCULATE YOUR AMPERE-HOUR NEEDS

Before buying batteries, you need to know how many amp-hours you want and at what voltage. This is normally accomplished by checking your load needs and deciding how many days your batteries must support this load when no charging power is available. For example, from your system, let's say we are operating:

100-watt kitchen light, 4 hours per day $100 \times 4 = 400$ watt-hours

60-watt bedroom light, 1 hour per day $\quad 60 \times 1 = 60$

60-watt bathroom light,
1 hour per day $\qquad\qquad 60 \times 1 = 60$

30-watt television (12V portable)
6 hours per day $\qquad\qquad 30 \times 6 = 180$

Total watt-hours per day $\qquad\qquad 700$

With a 12-volt system, $700 \div 12 = 58.33$ amp-hours per day. If you are expecting three successive days of calm winds (as is common in California) for a windmill input, then your battery requirements are:

58.33 amp-hours per day $\times$ 3 days $=$ 175 amp-hours

This amp-hour battery requirement at 12 volts can be met by two 12-volt, 87.5 amp-hour (or more) batteries in parallel, or one 12-volt battery rated at 175 amp-hours or more.

How often and how far you discharge your batteries affects the life of the battery. Battery manufacturers will tell you that no battery will last very long if completely discharged every time it is used. Planned discharge should not exceed 50% of the battery charge on a routine basis.

How much difference does it make whether you use auto batteries or deep discharge batteries? An auto battery will lose 20% of its original capacity after 100 to 150 complete discharges. A deep discharge battery will take 250 to 300 complete discharges before reaching the same point. Why the difference? Auto batteries are designed for high amperage starting loads (engine starts). In order to create that high current capability more plates are used and they are thinner. When deep discharges occur the plates can warp or shed active materials easily. Deep discharge batteries have thicker and fewer plates. They cannot support an extremely high current demand but are less easily damaged during deep discharge.

The warranty life of an auto battery assumes no deep discharges. Therefore, a 6 year-warranty-battery fully discharged 40 times per year may only last 3 years. If discharged to only 50% charge level, it might last the full warranty. Most auto battery warranties are not valid unless the battery is used in an auto, so don't count on free or prorated replacement. If your system will be deeply discharged often, use a deep discharge battery instead.

How do you know if the quantity of deep discharges warrants a deep discharge battery? See the sample battery lives below:

Qty of Deep Discharges Per Year	Auto Battery Life (But not above warranty yrs)	Deep Discharge Battery Life (But not above warranty yrs)
52 (once a wk)	2 to 3 yrs	5 to 6 yrs
24 (twice a mo)	4 to 6 yrs	above 6 yrs
12 (once a mo)	8 to 12 yrs	above 6 yrs

You might think—"I'll get a deep discharge battery to be sure I'm covered for a long time"—but look at the warranty periods available. Most deep discharge batteries are only warranted for 2 years; some are less. I know of one at 3 years (DEKA 8085, 85 AH, 12 V, $68). The warranty assumes perhaps 100 deep discharges per year. If less abuse of the battery occurs, it could last several years longer. But there is no guarantee. Your home electric system has to be in real trouble before you really need a deep discharge battery. Normally, you must expect your energy sources to keep your batteries near full charge.

If you use auto batteries, doubling your needed discharge capacity is recommended to reach maximum warranty life *only* if discharge of batteries occurs more than once a week. If discharging of batteries is expected only once a week or less, then auto batteries are recommended with no extra storage needed.

MAINTENANCE AFFECTS WHAT YOU BUY

Maintenance affects what batteries you buy. I had considered a lifetime auto battery that had no water replacement needs, but this had no way of checking cell charge with a hydrometer. For batteries in groups, it is essential that individual cells be checked for routine maintenance and for trouble-shooting battery problems. Although a voltmeter with spike probes can give cell voltage status it's not quite as good as the hydrometer. The batteries you buy will have a warranty life. The actual life will equal or exceed the warranty life up to 50%. For example a 2-year warranty battery will last 2 to 3 years. You can expect failure anytime after the warranty has expired. In general, if you plan to use your system for many years, the best buy on batteries is any battery with a 4-year or longer warranty. This gives you the most for your money.

HOW TO SELECT YOUR BATTERIES

Auto, truck, or golf cart batteries—which to buy? Golf cart batteries are designed for continuous, deep discharge, however, I find auto batteries stand up well for home electric systems even when fully discharged once per week. The advantage to truck batteries is higher amp-hours in a single package, however, the amp-hour rating can be matched by paralleling auto batteries. (It should be noted that a cell shorting in a parallel battery operation will cause very high currents in the good batteries until completely discharged. Therefore, some consideration of using series battery systems only should always be kept in mind while planning your storage needs.) Since any of these batteries will do for a home electric system, cost will have to determine the choice. Let's compare the four types for a 12-volt system requiring 200 amp-hours (AH).

As you can see from Figure 7-2, automobile batteries are the cheapest way to obtain the desired capability. Furthermore, additions to the battery system using automobile batteries can be accomplished in $46 increments whereas with the others, the additional increments are over $75 for each addition. From my experiments, I have concluded that automobile batteries come closest to being designed correctly for home electric systems. Furthermore,

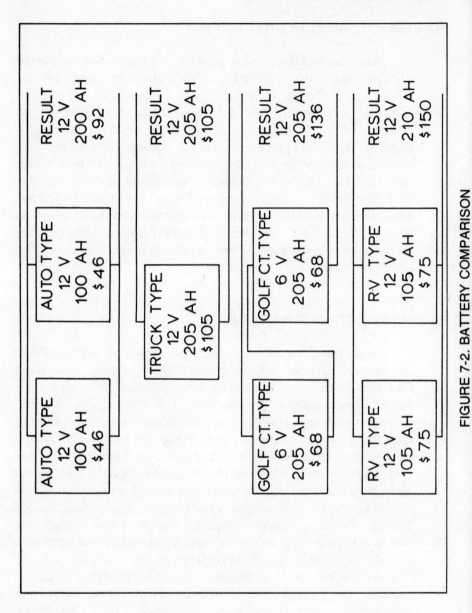

FIGURE 7-2. BATTERY COMPARISON

their cost and availability are excellent because of the high-volume market. Reliability is usually superb. You have only to take reasonable precautions in their use as discussed in this chapter.

HOW TO CONNECT GROUPS OF BATTERIES

Assuming that the 12-volt, 100AH auto battery is our standard unit, the following connection methods are used for various system voltages. (See Figures 7-3, 7-4, and 7-5.) In Figure 7-3, you will notice that batteries added in parallel increase the amp-hour rating of the total storage without increasing the voltage output. As many batteries as desired can be added regardless of their amp-hour ratings. The total amp-hour rating is the sum of the individual battery amp-hour ratings. All batteries connected in parallel *must* be of the same *voltage* rating. The buses shown are usually made from bars of copper or aluminum (steel or iron may be used) with attachments being bolted into tapped holes in the bus. Connections from the batteries to the buses are then accomplished with standard auto battery cables which can be purchased in many, pre-cut lengths.

In Figure 7-4, a method of charging at 24 to 28 volts when 12-volt loads are used is shown. The main reason this is handy is because of generator utilization. Let's say you have a 28-volt generator with an armature current rating of 3 amps continuous. That means the power output of the generator can be 84 watts (28×3). If you use this for charging 12-volt batteries in parallel at 14 volts, when you reach the maximum current rating of 3 amps you will only be putting out 42 watts (½ capability). In order to use the full capacity of your generator, you are forced to operate at rated voltage.

Further expansion of this idea is possible. Let's say you have a 120-volt, PM field, DC motor with a 3-amp maximum continuous armature current rating. That's a ¼ HP unit (about 200 watts). If you use it to charge a 12-volt battery system from a water wheel, you will have to adjust the motor RPM to put out 3 amps maximum (42 watts at 14-volt charge level). You are using less than ¼ of your capability. Assuming that you have enough water to get the full amount of power, you could charge at the 48-volt level across four 12-volt batteries in series. Then you would split your 12-volt loads

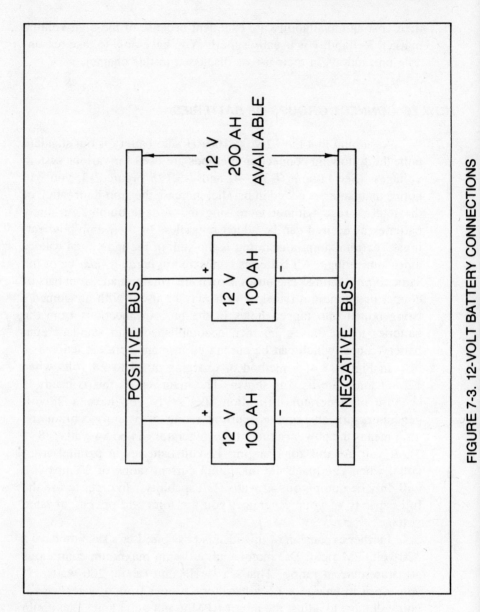

FIGURE 7-3. 12-VOLT BATTERY CONNECTIONS

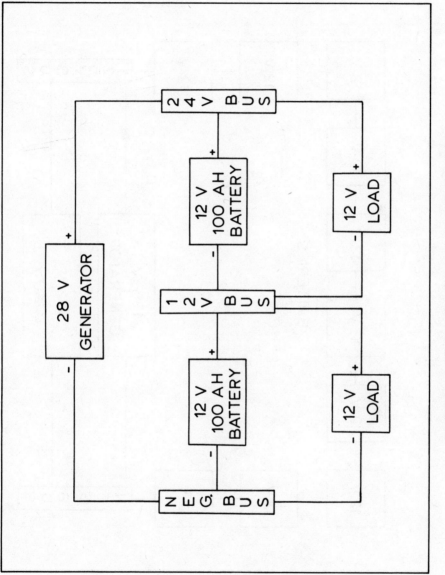

FIGURE 7-4. 24-VOLT SOURCE, 12-VOLT LOADS

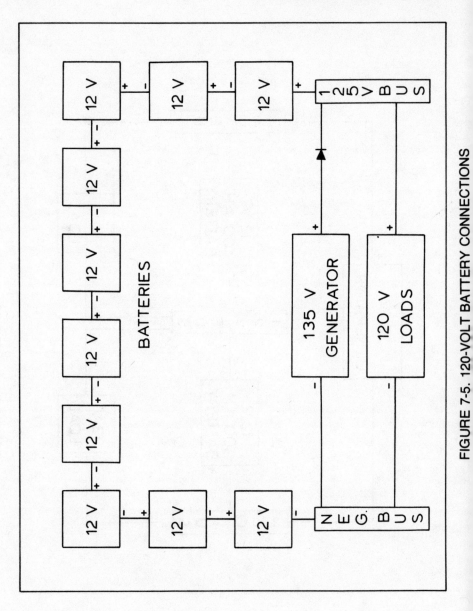

FIGURE 7-5. 120-VOLT BATTERY CONNECTIONS

up, ¼ across each battery. Your power input would be 168 watts (56-volt charge voltage at 3 amps) instead of 42 watts. The gear ratio and generator RPM must change whenever you change voltage outputs if you are to maintain the same current output. Remember when doing this that batteries in *series must* have the *same amp-hour rating* and batteries in *parallel must* have the *same voltage rating*.

The only other way to get in trouble is an uneven load on batteries in series. For example, if two 12-volt batteries are placed in series (charged by a 28-volt generator) and most of the 12-volt load is on one battery, it will result in one battery being at low charge and the other at high charge. Attempts by the 28-volt generator to charge these in series will result in over-charge of the high-charged battery and undercharge of the low-charged battery. During overcharging a battery will boil off the water, forming gases in the battery compartment and thus reducing battery life. During undercharging of a battery, the battery simply will not reach full charge and load demands to carry on for a prescribed period will result in a dead battery before the amp-hour rating is used. Therefore, even loads on batteries used in this manner are mandatory.

In Figure 7-5, you will notice that it takes ten 12-volt batteries in series to obtain a 120-volt DC system. This arrangement is referred to as a 125-volt DC system (nominal voltage) and is used often in commercial control schemes. Since it is still direct current (DC) the loads that you use are restricted. Loads that will operate with either AC or DC at 120 volts are incandescent lights and heater elements. The following devices will not work on 120-volt DC when they have been designed for 120 VAC use: dimmer switches, fluorescent lights, fans, blowers, transformer-input devices, motors, and compressors (such as refrigerators and air conditioners). A common use of 120-volt DC power in the home is for an existing lighting branch circuit. This branch circuit *must not* have any fluorescent lights or wall plugs in it. Furthermore, if the incandescent lights have dimmer switches on them, the dimmers must be replaced with toggle, on-off switches of the heavy duty type. (More on this in Chapter 8 on Effective Use of Private Power).

BATTERY ENVIRONMENT

Where should the batteries be and how should they be protected? Assuming a convenient location, the batteries should be halfway between the input sources (windmill, water wheel, solar cells, etc.) and the loads (lights, TV, heater, etc.). Do *not* put them on a *concrete* floor. The batteries should be well-ventilated, protected from extreme hot and cold temperatures, and kept dry. For example, under a porch in a dry place and protected from extreme temperatures by a wooden box with a hinged lid and ventilation holes is fine. If placed inside, as in a basement or under the house, make the wooden enclosure relatively airtight and attach two 1½" plastic pipes (or larger) to the outside for venting any gases off safely. You need access to the top of the batteries so make your door accordingly. Also remember these batteries must be kept upright at all times to prevent acid spillage. Therefore, don't put them on a shelf where they could be easily knocked off.

Rules and regulations for batteries are not abundant but they do provide safety guidelines. The National Electrical Code is your major rule book. Telephone company rules and state hospital codes are both more stringent for the installation of batteries. However, compliance with the National Electrical Code will assure reasonable safety for your home-use system. For your convenience, I will summarize the pertinent information from the National Electrical Code:

1. Article 480 of the National Electrical Code applies to all stationary storage battery installations using acid or alkali as the electrolyte with a nominal voltage in excess of 16 volts. Nominal voltage is defined as 2.0 volts per cell for lead-acid type batteries and 1.2 volts per cell for alkali type.
2. The code includes insulation requirements for the batteries that manufacturers of the batteries must comply with.
3. Racks or trays are required for support of batteries. Racks are defined as frames designed to support batteries or trays. Trays are defined as frames or shallow boxes. Racks must be substantial and made of treated wood or treated metal with nonconduct-

ing members directly supporting batteries or with suitable insulating material over the metal. The treating of wood and metal should be resistant to deteriorating action by the electrolyte. Trays should be resistant to electrolyte deterioration also, and made usually of wood or other nonconducting material.

4. Battery rooms or enclosures are required only for batteries in unsealed jars or tanks (such as most auto batteries which have water fill caps) where the total storage capacity exceeds 5 kilowatt hours. (This is calculated by the amp-hour rating times voltage equals watt hours. A lifetime 12-volt auto battery is worth about 100 amp-hours and therefore about 1200 watt hours or 1.2 kilowatt hours.) This means that if you have 5 or more batteries like this you are required to have a room or enclosure for them. If you have such a setup you are required to ventilate the gases from the batteries to prevent the accumulation of an explosive mixture in the battery room or enclosure. In storage battery rooms and enclosures, the wiring methods allowed are bare conductors, open wiring, Type MI cable, Type ALS cable, or conductors in rigid conduit or electrical metallic tubing. Conduit when used must be corrosion resistant material or corrosion protected material (i.e. PVC plastic conduit or galvanized metal conduit). Varnished-cambric-covered conductors, Type V shall *not* be used. Bare conductors shall *not* be taped. Where metal raceway or covering is used in the battery room or enclosure, it must stop at least 12 inches from the battery connection, be glaze insulator bushed, and be sealed to electrolyte entry by sealing compound, rubber tape, or other suitable material.

5. Article 503-15 of the National Electrical Code states that battery charging equipment shall be located in a separate enclosure (if in the battery room). Away from the battery environment, battery chargers do not have to be enclosed. Therefore, the author recommends that the battery charger *not* be located in the battery storage room or enclosure.

6. If you are considering putting batteries or battery chargers in an area where four or more gas-fueled vehicles are stored, you must comply with Article 511 which does not allow batteries, their chargers, or control equipment within 18 inches of the floor or below floor level.

Batteries are dangerous if proper connection and storage are not adhered to. Do not allow any metallic objects to come near to a position that would short the terminals (positive to negative)—not even for checking to see if voltage is there. Each auto battery has the capability of producing about 400 amps under short circuit conditions for a brief time. If four batteries are connected in parallel, 1600 amps will flow if any of the four batteries are shorted. These currents will burn or weld anything in their path so *never* allow yourself to become careless around these batteries. It is true that you can put one hand on each bus without shock hazard due to the low voltage (12-volt), but don't let that fact lull you into providing a decent electrical path for a short. Also with 120-volt battery systems, the shock hazard is as risky as 120 VAC so treat the live components with proper respect.

The next chapter will discuss the effective use of private power. Loads must be matched to battery voltage. If not matched, inverters or converters may be used to properly interface them.

8.

Effective Use of Private Power

SELECTING EFFICIENT LOADS

Select efficient loads? How? You cannot be efficient without knowing what the loads are, how much power they use, and how often they are used. Neither can you eliminate necessary loads. However, knowledge of loads and the latest efficiencies available can result in lower electricity use by trimming uncontrolled use of high-wattage devices and replacing some inefficient loads. Intelligent, cost-effective decisions are based on facts. Here are the facts. The loads shown in Figure 8-1 exclude lighting which will be discussed as a separate subject.

Looking at Figure 8-1, some conclusions can be drawn. A self-cleaning oven uses about 24¢ more electricity per month than one without self-cleaning. A frostless 15 cu. ft. freezer uses about $1.91 more electricity per month than one without automatic defrosting. A frostless 14 cu. ft. refrigerator uses about $2.32 more electricity per month than one without automatic defrosting. An automatic dishwasher costs about $1.20 per month electricity to operate. A clothes dryer costs about $3.26 per month electricity to operate. If you have a quick recovery electric hot water heater it is costing you $2.00 per month more electricity than one without quick recovery elements. Air conditioning costs between $10 and $30 per month depending on the area cooled (summer only). These comments are not meant to cause you to rush out and buy new equipment that operate at lower electric usage levels. That is not economical. But when your high power devices wear out, consider non-

APPLIANCE	AVG. WATTS	AVG. USE (Hrs. per month)	KILOWATT-HOURS USED	MONTHLY COST @ .4¢/KWH
COOKING				
Blender	385	3	1	$ .04
Broiler	1400	6	8	.33
Coffee Maker	900	10	9	.36
Deep Fryer	1400	5	7	.28
Frying Pan	1200	13	16	.62
Mixer	125	9	1	.04
Microwave Oven	1500	11	16	.66
Range w/oven	12,000	8	96	3.84
Range w/self-cleaning oven	12,000	8.5	102	4.08
Sandwich Grill	1200	2	2	.09
Toaster	1100	3	3	.13
Waffle Iron	1100	2	2	.09
FOOD STORAGE				
Freezer (15 cu.ft.)	340	292	99	3.97
Freezer (Frostless) (15 cu.ft.)	440	334	147	5.88
Refrigerator (12 cu.ft.)	240	252	60	2.42
Refrigerator (Frostless 12 cu.ft.)	320	316	101	4.04
Refrigerator (14 cu.ft.)	325	291	95	3.78
Refrigerator (Frostless 14 cu.ft.)	615	248	152	6.10
HOUSE ACCESSORIES				
Clock	2	730	1.5	.06
Floor Polisher	305	4	1	.05
Sewing Machine	75	12	1	.04
Vacuum Cleaner	630	6	4	.15
Trash Compactor	400	2	1	.03
Garbage Disposal	450	6	3	.11
WASHING				
Dishwasher	1200	25	30	1.20
Clothes Dryer	4800	17	82	3.26
Hand Iron	1000	12	12	.48
Washing Machine	500	17	9	.34
Water Heater	2500	142	355	14.20
Water Heater, w/ quick recovery	4500	90	405	16.20

FIGURE 8-1. TABLE OF ELECTRICAL LOADS

APPLIANCE	AVG. WATTS	AVG. USE (Hrs. per month)	KILOWATT-HOURS USED	MONTHLY COST @ .4¢/KWH
COMFORT				
Room Air Conditioner (summer)	900	300	270	10.80
House Air Conditioner (summer)	4800	150	720	28.80
Electric Blanket (winter)	175	120	21	.84
Attic Fan (summer)	375	300	112	4.50
Circulating Fan (summer)	90	300	27	1.08
Hair Dryer	380	4	1.5	.06
Portable Heater (winter)	1300	50	65	2.60
ENTERTAINMENT				
Radio	70	100	7	.28
B & W TV, Tube type	160	182	29	1.16
B & W TV, Solid state	55	182	10	.40
Color TV, Tube type	300	182	55	2.18
Color TV, Solid state	200	182	36	1.46

FIGURE 8-1 Continued

self-cleaning, non-frostless, non-quick-recovery units for replacement. Weigh your decision carefully against monthly costs—for some, the cost may be worth the advantages. Each situation may require a different decision.

LIGHTING

Lighting varies so drastically from home to home that no realistic cost level is available for electricity. However, my own light usage averages about 120 kilowatt-hours per month in an 1800 sq. ft. house with 4 people. At 4¢ per KWH, that's $4.80 per month. It is higher in the winter and lower in the summer. The only fluorescent lights I have are 240 watts recessed into the kitchen ceiling. The balance are incandescent lamps. Lighting represents 10% of my total electrical load.

A few years ago, the Federal Energy Office (now the Federal Energy Administration) called for a reduction of lighting in all commercial and industrial buildings. The nonmandatory guidelines specified lighting levels of 50 footcandles at work stations, 30 footcandles for general work and sales areas, and 10 to 15 footcandles in hallways and corridors. Before this awareness of reasonable levels that would conserve energy, many illumination areas in commercial and industrial buildings exceeded 100 footcandles! Many such areas removed every other fluorescent bulb to reduce electrical usage. The lighting industry responded by selling lamps that were more efficient, i.e., more lumens per watt input. This efficiency in lighting is called *efficacy* and is defined as the lumens of light created per watt of power used. Figure 8-2 compares lamp types.

The efficacy on incandescent lamps is lowest. (If the lamp comes on immediately and is not a tubular fluorescent, then it is an incandescent lamp.) Incandescent lamps give off 50% or more of their energy in heat instead of light. It will help heat your house in winter, but in summer, it fights the air conditioner. Therefore, the reaction by individuals is to use more fluorescent fixtures. A fluorescent lamp puts out almost twice the light of an incandescent while using the same electrical power and lasts about 3 times as long before lamp replacement is required. The initial cost of fluorescent fixtures is higher than incandescent types, but I would not call them expensive.

The high intensity discharge (HID) lamps are mercury-vapor, metal-halide, and high pressure sodium. (A tungsten-hologen lamp is an incandescent.) In general, HID lamps put out more light, cost more, and are cheaper to operate. Their life is comparable to fluorescent lamps, but cost lower to operate because less lamps are required to produce the same light levels. Mercury-vapor lamps are less efficient than fluorescent. Both metal-halide and high-pressure sodium are more efficient than fluorescent with high-pressure sodium being twice as efficient as fluorescent and four times as efficient as incandescent. I would not recommend HID lights inside of the home, but for outside use (especially where a light is left on

CHARACTERISTICS	INCANDESCENT	FLUORESCENT	MER-CURY VAPOR	METAL-HALIDE	HIGH PRESSURE SODIUM
Lamp Wattages	15 to 1500	40 to 219	40 to 1000	400, 1000,1500	75,150, 250,400, 1000
Life (hrs)	750-12,000	9,000-30,000	16,000-24,000	1500-15,000	10,000-20,000
Efficacy (lumens/watt)	15 to 25	55 to 88	20 to 63	80-100	100-130
Relight time	Immediate	Immediate	3 to 5 min.	10 to 20 min.	less than 1 min.
Comparative fixture cost	Low	Moderate	Higher	Higher yet	Highest
Comparative operating cost	High	Moderate	Lower	Lower yet	Lowest

FIGURE 8-2. LAMP TYPES

all night, consider a high-pressure sodium light of 75 watts; a 100-watt unit is also available.) It will put out light equivalent to a 300-watt incandescent lamp without running up the electric bill nearly as much. The sodium light is orange.

SAVING MONEY ON LIGHTING

As incandescent fixtures get old, I recommend their replacement with fluorescent fixtures. However, if fluorescent lighting is placed over a face make-up area, be sure to use warm-colored lamps or have some incandescent ones also, otherwise colors seen in the mirror will be different than in daylight. When lighting a kitchen area, it is recommended that two or three switches be provided to change the amount of lighting used. Preparing a meal needs more light than a goody run during TV commercials.

Incandescent lighting can be run on DC battery voltage if the correct voltage is used. Your existing incandescent lamps in the house can be connected to a 120-VDC battery bank. If you would like to use 12 volts on your lighting without changing the fixtures, 12-volt bulbs for use with the same base are available in 25-watt and 50-watt sizes (slightly over $1 each) at camper accessory stores. All loads must be designed for the battery voltage to which they are connected or for the inverter output if one is used.

Lights should be your first DC load. Did you ever notice that many lights are placed on ceilings? These are inefficient at providing light for eating and reading. For example, in my dining room I had a chandelier with five 60-watt bulbs in it. The bulbs shone upward and much of the light never reached the table. Now I have attached five 15-watt, 12-volt lights to the bottom of the chandelier with focused lenses (auto back-up lights). I get twice the light on the table and use only ¼ the power. Thus for every kilowatt-hour of electricity I use there, I save 4 kilowatt-hours on my electric bill. Furthermore, each of the five lights cost under $3 each. I replaced the 60-watt, 120 VAC bulbs with 7-watt flickering bulbs (simulating candles) so now I can eat by simulated candlelight occasionally. The chandelier is more functional than before and is still beautiful. (See Figure 8-3.)

FIGURE 8-3. DINING ROOM CHANDELIER MODIFICATION

This approach to lighting may be used throughout the house if desired. Parts of your home that require area lighting such as a kitchen or possibly a large pool table should be illuminated with fluorescents. However, where you eat, sleep, read, or watch TV you could install a 12-volt directional lamp. (This lamp should be placed 2 to 3 feet from the reading surface.) In my home, I found 30 locations where such lamps could be beneficially installed: on the wall over each bed; on the hallway wall; on the wall over each desk; over each sink and toilet; suspended from a ceiling post or table lamp located strategically along the sofa and near chairs; over each place at the breakfast bar; and over the work bench. Remember these lamps can be made beautiful as well as functional. Each light should have an individual switch on it that can be reached from the using position. In some areas you may wish to connect wall switches to the lamps so that you can enter and exit the room in light. Some wall lamps are available with an 18-watt bulb at under $7 each. Beautiful brass spotlights are available for under $15. With some thirty 12-volt lighting fixtures in my home, the only place I use commercial-powered lighting is in the kitchen. A check on the original lights normally on in the house at night indicated that I used 4 kilowatt-hours (KWH) per day of commercial electricity which is now replaced by 1 KWH of battery electricity. Since I used to use 40 KWH per day of electricity on commercial power before I built my system, the lighting alone is a 7.5% savings on my electric bill. This percentage may vary for others. (My 36 KWH per day use of electricity other than lighting comes from a refrigerator, dryer, electric stove, microwave oven, central heating and air conditioning fans, swimming pool pumps, and plug-in appliances.)

AVAILABLE 12-VOLT LIGHTING FIXTURES

Many types of 12-volt lighting fixtures are available. (See Figure 8-4.) If you can find a camper, trailer or recreation vehicle accessory store in your area, brouse through the lighting section. You should find fluorescent 12-volt fixtures for ceilings and walls rated at 8 watts and 15 watts per lamp. These will draw between one

(Photo courtesy KW Industries, Citrus Heights, Ca.)

FIGURE 8-4. 12-VOLT LIGHTING FIXTURES

and two amperes and put out more light than an incandescent lamp of the same wattage rating. 12-volt fluorescent fixtures cost about $22 for a double 15-watt fixture and may be cost-effective in some locations where this fluorescent fixture could replace 60 watts of incandescent lamps and use half the power doing it.

SAVING MONEY ON 12-VOLT APPLIANCE LOADS

Television

A good 12-volt load is television. Not because you use a lot of electricity here but because it is a source of entertainment to be used when commercial power failures occur. An average 25-inch solid state color TV uses about 250 watts. This would consume one KWH of electricity per average viewing day. If it is a tube type, it could draw nearly twice the power. I know of no 25-inch TV's that are made for 12-volt operation, but many 12-volt portable TV's are available, both black and white and color. You may already have one. Run this portable TV on your 12-volt electric system.

What about the cost of that 12-volt DC TV? Sears and Roebuck has a 12″ black and white TV that has both 120 VAC or 12 VDC operational capability. (A 12″ screen can be viewed comfortably from about 10 feet.) The cost of this one was about $130. On 12 volts DC it uses 16 watts; on 120 VAC it uses 33 watts! Hitachi markets a 12-volt 9″ color TV. (A 9″ screen can be viewed comfortably from about five feet.) The cost of this one was $320. On 12-volts DC it uses 30 watts; on 120 volts AC it uses 60 watts. Larger TV's are not currently made for 12-volt operation, but more 12-volt TV's should be available in the future. The two I mentioned are reasonable examples of today's market. (There are many black and white DC TV's available, but only two color DC TV's.)

To be quite honest, a TV is a luxury, especially a second one. However, it is probably the cheapest form of entertainment available today. As such, whether you elect to use a second TV on your 12-volt DC electric system or replace your wornout TV with one that will operate on 12 volts; you will be saving electricity and providing an entertainment source. If you replace a 25-inch color

TV with a 9-inch 12-volt color TV, you would use 0.12 KWH per day from your batteries instead of 1 KWH from your commercial power! This would take approximately a dollar off your monthly electric bill.

Cooling Fans

Fans can be obtained that run on 12-volt DC for about $11. You find them in recreation vehicle accessory stores. These are small and suitable for blowing on one or two people.

Why use fans in an air conditioned home? Simply because it can save considerable electricity. If you normally run your air conditioning at 70°F, consider running it at 78°F and use small, 12-volt fans at the places where you work or play to obtain equivalent body cooling. A change in the thermostat setting of 8°F should reduce the electricity used by at least 10% depending on outside air temperature and house insulation. In my house such savings would represent about 2 KWH per day. This alone could save 5% of your electric bill in the summer if you use air conditioning. If you don't have air conditioning, substitute several small 12-volt fans for your large table fans. This will save a small amount of money on your electric bill.

Heating

Portable heaters can be used to allow reduced central heating temperatures. Similar savings to air conditioners can be obtained for your heating system. The idea behind a portable heater is that it provides individual heat so that your central heating can be turned down and you can still remain comfortable.

When you get cold, the extremities are chilled (ears, nose, hands, and feet) and the easiest and most effective extremity to heat is your feet. I did it with incandescent lamps; they put out a lot of heat. (See Figure 8-5.) The foot-warmer is extremely simple to make. It is a cylindrical-shaped foot stool with a hollow center. Over the top is affixed a net. About 4 inches from the bottom of the footstool,

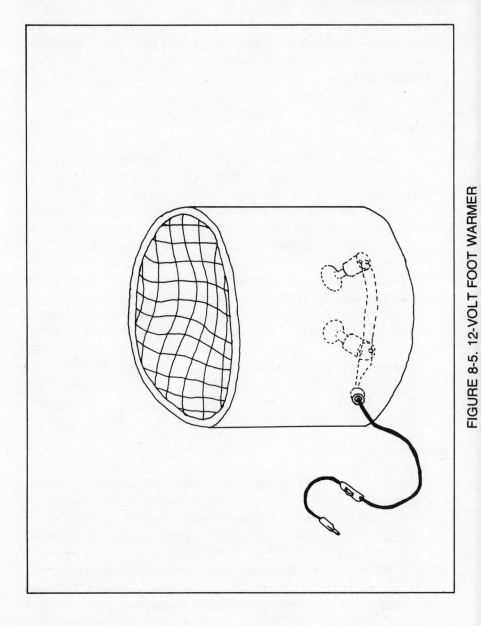

FIGURE 8-5. 12-VOLT FOOT WARMER

mount several 12-volt incandescent light bulbs facing inward. Make the electric cord long enough to reach your 12-volt wall receptacle. Use a cigarette-lighter plug. Put an in-line switch where you can reach it without moving from your sitting position. Put your feet on the stool net, turn it on, and cover your feet with a blanket or towel. When you get too hot, turn it off. It's simple and effective. This is not a UL-approved device so you may want to buy a 120 VAC portable heater and use it locally from commercial power. You will still save on your electric bill by running central heat at lower temperatures.

Other Loads

There is no limit to 12-volt appliances that you can run from your battery system if you can generate enough power. Take a look in a catalog of accessories for mobile homes, recreation vehicles and camping. You will find 12-volt intercom sets, 12-volt vacuum cleaners, 12-volt hand spotlights, 12-volt water pumps, 12-volt air compressors, 12-volt two-speed fans, 12-volt fluorescent lights, 12-volt incandescent lights, 12-volt refrigerators, 12-volt inverters, 12-volt tape players, 12-volt AM-FM radios, and 12-volt battery protectors. The refrigerators I saw used 42 watts or 60 watts and were over 2 cu. ft. in size.

The 12-volt battery protector is a device normally used in dual battery systems. It allows both batteries that are connected in parallel to be charged by the charging generator but loads can only discharge one of them. This keeps one battery from being discharged by loads on the other battery thus allowing one battery to be saved for a certain function such as starting the car or, as in our case, for some specific emergency function such as a fire alarm or burglar alarm.

You will notice that in each use of your 12-volt power we have strived to get the best efficiency from the load. This should be your constant goal. Some basic rules that will help are:

Lighting
1. Always provide just enough light to do the job needed.
2. Always choose fluorescent lamps first, if they will meet your

needs, because they provide more light per watt than incandescent lamps. Their best use is where they can suffice for two or
three incandescent lamps.

Television
1. Black and white: buy a 12-volt TV before considering an inver-
    ter to run an existing 120 VAC TV. It's cheaper!
2. Color and B & W: Always use your dual-powered 12-volt TV
    on 12-volts if it is available. It will draw less than 30 watts from
    batteries as opposed to twice as much on commercial power.

Fans
1. Always use your 12-volt fans so that you can run your air
    conditioner at 78°F (adjust temperature for comfort in your
    humidity level).

Heaters
1. Always use your individual heater so that you can run your
    **commercially fueled heater at 65°F (adjust for comfort as low as
    possible).**

Other 12-volt Appliances
1. When adding new appliances, try to obtain a 12-volt one if you
    have the battery power available. But remember that you must
    increase battery quantity and charging systems before increasing
    loads beyond your existing charging capability.
2. Consider your 12-volt system as an emergency power source for
    fire alarms and burglar alarms. Add a battery for this function
    and put a battery protector on it so that it can't be discharged by
    other loads.

Should you have a 120-volt battery system, you can run incandescent lighting (now on 120 VAC commercial power) without
changing the fixtures by removing the AC connection at its circuit
breaker and connecting to your batteries through a fuse size of 15
amps for #14 house wiring or 20 amps for #12 house wiring. In
most cases this will be the same size fuse as the circuit breaker at the
removed connection. The negative of your battery system can be
attached to neutral of the AC system. Be sure it is a grounded
neutral and no other connections to AC take place. Be sure that no
120 VAC fluorescent fixtures exist on DC circuits. If dimmer

switches were previously employed on the incandescent fixtures now to be used on DC battery power, you must replace the dimmers with on-off switches as the dimmers won't work on DC battery power and could be damaged.

HOW TO PROVIDE 120 VAC FROM YOUR BATTERIES

You may elect to run existing 120 VAC loads from your battery system. To do this you will need an inverter that converts your DC battery voltage to 120-volts AC. Two general types of inverters are available: those that produce square wave output and those that produce sine wave output. Standard house current is a sine wave and all 120 VAC equipment will run on the sine wave. The square-wave-output inverter is cheaper to buy but won't properly run a color TV, refrigerator, air conditioner, or freezer. The square-wave-output inverters will run lights, heating elements (such as hot water), electric shavers, toasters, and many small appliances. A 12-volt DC to 120-volt AC square wave inverter putting out 220 watts costs $75. Square wave inverters are available up to 500 watts. Sine wave inverters are of two types: solid state, and motor-generator. Motor-generator sets are available from 500 watts through 900 watts depending on input battery voltage from Redi-Line (Applied Motors, Inc., 4801 Boeing Dr., P.O. Box 106, Rockford, Ill. 61105). Solid-state inverters producing both square wave or sine wave can be obtained from Lombard's Lafayette electronic stores nationwide. Sine wave inverters are available at 250 watts, 500 watts, and 1000 watts output. A 12-volt DC to 120-volt AC sine wave solid-state inverter rated at 1000 watts continuous, 1100 watts peak power costs $300. Solid-state inverters are more efficient and are recommended. Ninety-five percent efficiency is not uncommon for these.

If you need more than 1000 watts of 120 volt AC from an inverter, use another inverter on a separately wired circuit. You can load them from the same battery circuit. *Don't* connect the 120 VAC outputs together.

If you should decide to power your whole house from a battery system, use 120 VDC (125 VDC system) batteries and a 5 kilowatt

(KW) or larger inverter with 220/110 VAC single phase output. Conventional loads can then be used. A heavily insulated solar heated and cooled house (1500 sq. ft.) could get by on a 5 KW inverter. If the house has conventional insulation, conventional heating and cooling, and about 2,000 sq. ft. of floor, it may require a 20 KW inverter. The 5 KW unit is practical. The 20 KW unit is extremely expensive and a fantastic amount of input power generation is required.

9.

How to Monitor and Protect Your Energy System

Monitoring your system is a natural desire. You want to see what you have accomplished and how it is performing on a regular basis. If you don't install permanent voltmeters and ammeters, then you will need to drag out portable equipment for performance checks. Even commercial power sources provide permanent metering at their monitor points; they don't have to be expensive. In addition to discussing monitoring, wiring and protection will be discussed.

WHERE TO MONITOR VOLTAGE.

Voltage is handy to know at many locations. Therefore, a selector switch is employed with a single voltmeter. (See Figure 9-1 for example connections.) Since all input devices have a common negative on the battery, your voltmeter (−) is connected to the battery negative. The positive side of the voltmeter is connected to one of several possibilities through a selector switch. This selector switch may be any available switch since the current flow will be one milliamp or less depending on the meter. (An "off" position is shown on the selector switch, but this is optional because the current draw is so low.) The other connections are made to the positive lead of each generating device between the diode and the generation unit. When more than one device is used to charge batteries, a diode in each positive input line is necessary. Also anytime a DC motor is

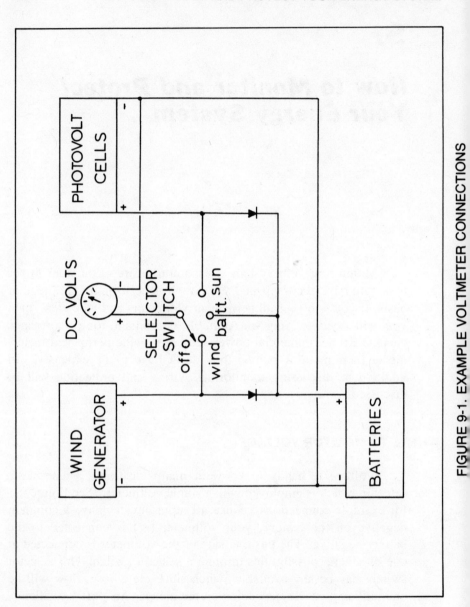

FIGURE 9-1. EXAMPLE VOLTMETER CONNECTIONS

used as a generator, a diode is required to prevent the unit from running as a motor.

A 0 to 15-volt DC voltmeter is adequate for a 12-volt system. For under $10 you can purchase one with 5% accuracy. With the connections as shown in Figure 9-1, the voltmeter will read battery voltage when battery is selected. It will read 0.6 of a volt higher when reading wind or sun input if these units are charging the batteries. If these are not charging the batteries, it will read whatever voltage is being generated. The diodes allow correct wind generator or photovoltaic cell voltage readings at all times.

HOW TO MONITOR CURRENT

Measuring amperage is a bit more complicated than measuring voltage. The ammeter is an in-line device and is difficult to switch. If the ammeter has no external shunt, it may be used in one place only. You may use the ammeter at a remote location if you bring both load carrying wires to that point. If you have an external shunt on your ammeter, say a 50 millivolt one, you may place the shunt in the line to be monitored and run smaller wires to the ammeter. Furthermore, you may place a similar shunt in each line to be monitored and switch both wires from the ammeter to the desired shunt for readout. (See Figure 9-2 for an example of shunt switching.) One disadvantage to shunt switching is a single range on the meter must be good for all shunt readings. Some range scaling can be done, but I will not go into that. I prefer one ammeter for each positive line with an appropriate scale for the device being monitored. This can be done with or without external shunts. Ammeters must be used with shunts if designed for them and without shunts if designed for direct in-line readings. (See Figure 9-3 for a sample wiring.)

PROTECTING WIRING WITH FUSES

Wiring should be sized for the amperage expected. Figure 9-4 is a table of wire sizes versus amperage ratings. Notice that current ratings for the same size wire vary with the insulation's capability to

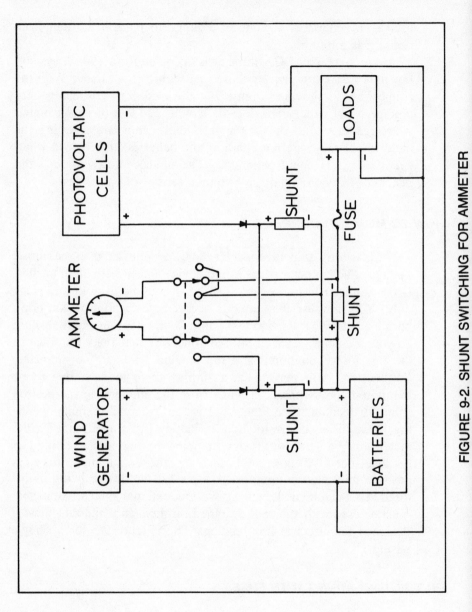

FIGURE 9-2. SHUNT SWITCHING FOR AMMETER

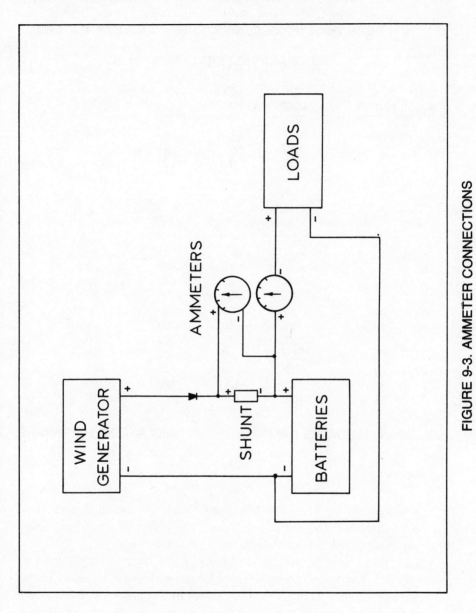

FIGURE 9-3. AMMETER CONNECTIONS

NOT MORE THAN THREE CONDUCTORS IN CABLE OR RACEWAY

TYPES AND TEMPERATURE RATINGS

SIZE AWG*	60°C (140°F)	75°C (167°F)	85°C (185°F)	90°C (194°F)	90°C (194°F)
	RUW (14-2), T, TW, UF	RH RHW, RUH, (14-2), THW, THWN, XHHW, USE	V, MI,	TA, TBS,SA, AVB,SIS	FEP,FEPB, RHH,THHN, XHHN(Dry only)
18	—	—	—	21	21
16	—	—	22	22	22
14	15	15	25	25	15
12	20	20	30	30	20
10	30	30	40	40	30
8	40	45	50	50	50
6	55	65	70	70	70
4	70	85	90	90	90
3	80	100	105	105	105
2	95	115	120	120	120
1	110	130	140	140	140

FIGURE 9-4. ALLOWABLE AMPACITIES OF INSULATED COPPER WIRE

withstand heat. The letter designations are insulation codes. They are important. *Do not disregard the capability of your particular wire—little margin exists and fire can result!*

Fuses instead of circuit breakers are recommended for DC circuits. The fuses should be sized for the maximum ampacity of the wire or smaller. If the armature current in a generator is rated lower than the wire, you may wish to size the fuse for the maximum armature current on that input circuit. The fuse will then prevent armature burn out. Each load circuit should have a fuse as shown in Figure 9-5. The fuses should be labeled by load. If numbered, provide a key list nearby as is done on circuit breaker panels.

SINGLE CONDUCTOR IN FREE AIR

TYPES AND TEMPERATURE RATINGS

	RUW(14-2) T,TW,	RH,RHW RUH(14-2), THW, THWN, XHHW	V, MI	TA,TBS, SA,AVB, SIS	FEP FEPB, RHH, THHN, XHHW(dry only)
18	—	—	—	25	25
16	—	—	27	27	27
14	20	20	30	30	20
12	25	25	40	40	25
10	40	40	55	55	40
8	55	65	70	70	70
6	80	95	100	100	100
4	105	125	135	135	135
3	120	145	155	155	155
2	140	170	180	180	180
1	165	195	210	210	210

*AWG—American Wire Gauge

FIGURE 9-4. Continued

AUTOMATIC ALARMING

You may desire to know when your battery is fully discharged—at that fine point where motor loads could be damaged or battery life shortened if you continue to operate. CALEX makes a battery discharge indicator (UL approved) Model 3655 with adjustable trip point for 12 to 48-volt batteries (price $98). (Contact CALEX, P.O. Box 555, Alamo, Calif. 94507.) Obviously, the price must be justified by loads that could be damaged at low voltage. If your loads are just lighting, the dimming lights will tell you if your batteries have discharged.

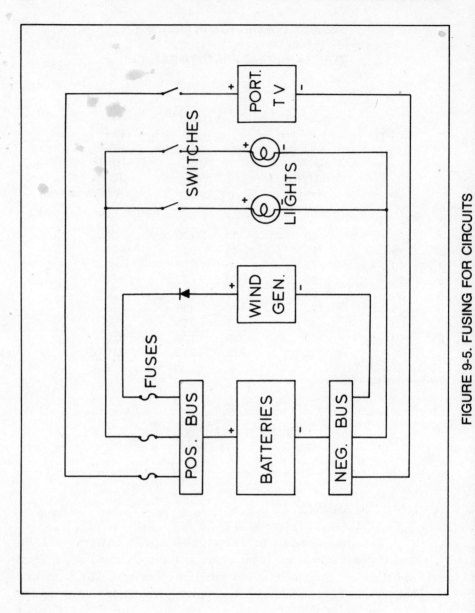

FIGURE 9-5. FUSING FOR CIRCUITS

You may wish to know when your batteries are overcharging. The Figure 9-6 circuit can be used to alarm either visually or audibly. This circuit can be used for any system voltage below 100 volts. The potentiometer is adjusted to trip at the desired voltage (15.0 volts for 12-volt systems).

USING CONTROL PANELS

Control panels are used to house voltmeters, ammeters, fuses, alarm devices, and buses. For 12-volt to 28-volt systems, you could make your own from wood as I did. For higher than 28-volt systems, commercial steel panels should be used. A great variety are available in electrical supply stores. You will have to modify them to fit your meters. Be sure to place your panel in a dry location (inside is preferred). Use large enough wire to carry the whole load from the batteries to the panel and put a main fuse in this line.

Circuit-breaker panels can be used. Circuit breakers do not act as fast as fuses and should be rated for the DC voltage they will see. An AC circuit breaker of the same voltage rating as your DC circuit *cannot* be used!

The remaining chapters tell how to build wind generators, a water wheel, and an exerciser, and how to maintain and evaluate your system.

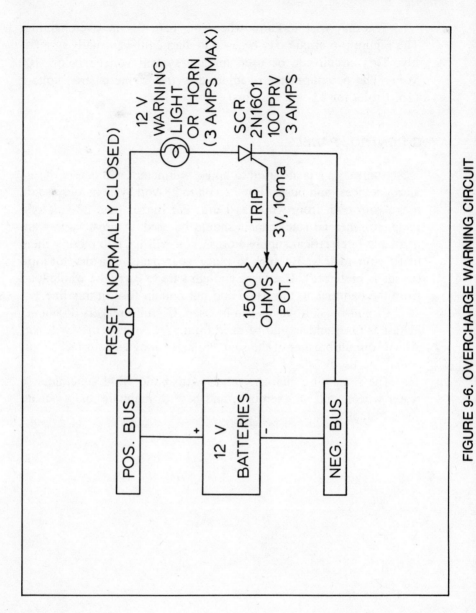

FIGURE 9-6. OVERCHARGE WARNING CIRCUIT

10.

Building the Vertical-Shaft Pierson Wind Turbine

This vertical-shaft wind turbine is a result of experiments by the author and is referred to as the Pierson Wind Turbine in Chapter 1. The unit started out as an S-rotor made from 55-gallon drums. These being unsuccessful in extracting power below 12-MPH winds, the system has evolved to its present configuration. Depending on wall length, this configuration will begin putting out power in 5 MPH winds or below (7 MPH without stators).

In Figure 10-1, the rotor is shown without its stator vanes. In Figure 10-2, the four stationary vectoring vanes, plastic covered walls, are in place. The main objective in proceeding with this approach (vertical shaft) was to achieve an omni-directional wind turbine capable of providing power at extremely low wind speeds without the need for a tower. The plain truth about height for a windmill is: Every bit you can get helps; particularly if you have nearby trees or building obstructions. The rule that a windmill should be 15 feet above all objects within 400 feet is still true for best results. Therefore, this vertical-shaft wind system is recommended only for roof-top or hill-top applications.

HEIGHT AND ZONING RESTRICTIONS

This wind turbine with its stationary walls can be considered a structure under building codes, although a building permit and inspection may not be required. Since it is a structure, building height

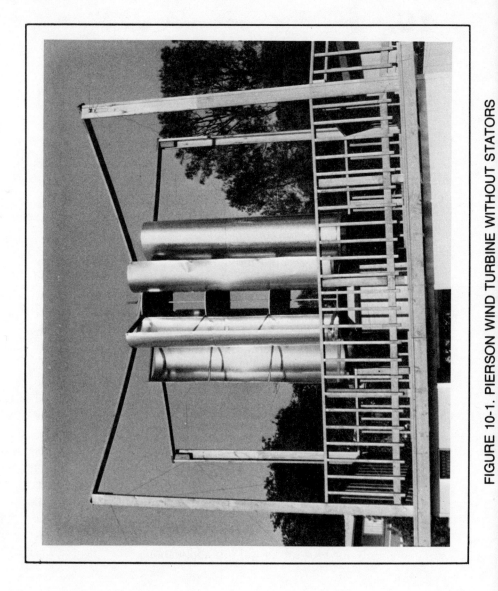

FIGURE 10-1. PIERSON WIND TURBINE WITHOUT STATORS

FIGURE 10-2. PIERSON WIND TURBINE WITH STATORS

and zoning restrictions will apply. For example, in my residential area, a 30-foot height restriction and nothing within 10 feet of the property line applies.

DETAILED DESCRIPTION

This turbine (the moving part) is 8 feet in diameter with 9-foot-high turbine blades. The eight blades are half circles, located 45 degrees apart around 4-foot diameter, ½-inch thick, plywood discs located each 3 feet of height. Each blade is made from aluminum sheeting 3 × 9 ft. or from 12-inch heater duct expanded to a 2 foot semicircle. The plywood discs are mounted on a 1-inch diameter solid steel shaft, 12 feet long. Bearings are used at the bottom and top to support the turbine. At the bottom, the bearing is attached to a box 3′ × 3′ × 3′ where the RPM is increased to drive a generator. The top of the turbine is supported via its bearing attached to cross-members that span 8½ feet between the stationary walls.

The four stationary walls are located 90° apart, are 12 feet high and a minimum of 8 feet long. These walls are used as if they were four square funnels feeding air to the turbine. If they extend out 8 ft., they will collect air from a 13 ft. wide × 12 ft. high area of 156 sq. ft. to feed to the exposed 36 sq. ft. of turbine. If walls 14 ft. long are used, they will pick up air from a 20 ft. wide × 12 ft. high area of 240 sq. ft. to feed to the 36 sq. ft. of exposed turbine. These 14 ft. long walls require a 20 ft. × 20 ft. roof or platform floor for turbine installation (typical 2-car garage size). Figure 10-3 is the parts list for this system. It assumes that a 20 ft. × 20 ft. platform exists.

TURBINE

Building Instructions

To start, build the turbine. Your first step is to cut out the four 4-foot diameter discs from ½-inch plywood. Use exterior plywood, smooth on one side (Type A-C). Plan your work so that the smooth side is installed upward. This will prevent rain water from puddling

QTY   **DESCRIPTION**

WALLS AND TOP SUPPORT

QTY	Description
(Your count)	2″ × 4″ × 12′ long to fit your wall plans
5 lbs	Nails, 10d
(Your count)	¼″ thick × 4′ × 8′ A-C plywood (to cover walls)
(Your count)	Paint for walls
2	1″ ball bearings, Browning FB 250-1 with 4 hole flange
4	⅜″ × 4½″ long hex bolts, with flat washer, lock washer, and nut. (for top bearing)
4	⅜″ × 2″ long hex head wood screws. (for bottom bearing) (pre-drill ¼″ holes)
(Your count)	2″ × 4″ bracing for ramps and roof
(Your count)	plywood cover for ramps and roof
2	2″ × 6″ × 10′ long (top bearing support)

RPM RATIO EQUIPMENT

QTY	Description
1	1″ thick A-C plywood, 24″ × 24″
2	½″ shafts × 7″ long, cold roll steel
4	½″ ball bearings with 4 hole flange, Sealmaster SF-8
1	12″ pulley for 1″ shaft
1	3″ pulley for generator shaft
8	$5/16$″ × 2½″ long hex head bolts with nuts

The following to be adjusted for desired RPM Ratio.

QTY	Description
1	_____ pulley for ½″ shaft
1	_____ pulley for ½″ shaft
1	_____ pulley for ½″ shaft
1	_____ pulley for ½″ shaft

GENERATOR

QTY	Description
1	DC Motor, PM field, rated 600 RPM, twice the battery voltage, and about 1 horsepower.

FIGURE 10-3. PARTS LIST FOR THE PIERSON WIND TURBINE

QTY   **DESCRIPTION**

TURBINE

QTY	DESCRIPTION
2	4′ × 8′ × ½″ thick A-C Plywood
½ gal.	Exterior Paint
1	1″ dia. × 12 ft. long cold roll steel shaft
2	1″ dia. pipe nipples × 4″ long (cut in half to make 4 pieces threaded on one end).
8	10-24 set screws
4	1″ pipe flanges
16	#10 × 1¼″ long machine screws with flat washer, lock washer, and nut (for flange to plywood installation).
32	2″ × 2″ × 3 ft. long redwood (turbine blade supports)
32	2″ × 4″ × 6″ long redwood blocks
128	¼″ dia. × 3″ long hex bolts
256	¼″ flat washers
128	¼″ lock washers
128	¼″ nuts
8	3′ × 9′ × .032″ thick aluminum sheets (or substitute 12″ dia. heater ducting).
64	$5/_{16}$″ × 2″ long hex head wood screws
64	$5/_{16}$″ fender washers
64	$5/_{16}$″ flat washers

BASE BOX

QTY	DESCRIPTION
1	3′ × 3′ × ¾″ thick A-C Plywood
2	3′ × 3′4″ × ¾″ thick A-C Plywood
4	4″ × 4″ × 3′ long fir beams
2	4″ × 4″ × 2′6½″ long fir beams
2	2″ × 6″ × 3′ long fir beams
4	3′ × 3′ sides (any available material) (one or two sides to be hinged for access)
2 or 4	hinges (for doors)
1 or 2	hasps or hooks (for doors)

FIGURE 10-3. Continued

in knot holes. (See Figure 10-4 for layout of the plywood.) Two sheets of plywood for a total of 4 discs are required. Locate the two centers by ruler and put a nail there (just start it). From the nail use string and pencil to mark the circles. Remove the nails and cut the discs out with a saber saw. Then drill a 1-inch diameter hole at the previously nailed center point. High-speed wood bits up to 1½ inch bit size are available for use in ¼-inch chuck electric drills. When using these bits, change the drilling to the back side as soon as the tip penetrates the back side of the wood to prevent splintering. Before mounting the plywood discs on the shaft, you will need to paint them and then mark the location of the eight supporting 2 × 2's as shown in Figure 10-5. (The discs are the only part of this turbine requiring paint and this is the easiest time to paint them. Use at least two coats of good exterior paint on both sides of each disc. When dry, proceed with marking.) In Figure 10-5, notice how the offset from the 45° lines is necessary so that the outer end of the 2 × 2 will be inside of the turbine blade for mounting purposes. Do this layout for each 2 × 2 on each disc.

Shaft preparation is important. The 1-inch diameter, cold-roll steel shaft should be 12-feet long. The ends must be slightly tapered on a grinding wheel for about ¾ of an inch to remove any bulges due to hot cutting of the length. If you don't have a grinder, have this done where you buy the shaft.

Before mounting the shaft with flanges and nipples, prepare the nipples as follows. Drill at least 2 holes in the nipple and tap it for any available size of screws to be used as set screws. I used 10-24 size with small holes in the head for safety wiring. When this is done, slide the 1-inch diameter shaft through each disc and its flange and nipple so that each flange and nipple can be installed as shown in Figures 10-6 and 10-7. Be sure to align a 45-degree mark on each disc in a vertical line parallel to the shaft so that your turbine blades will be vertical when installed. Tighten the nipple into the flange, slide it on the shaft up against the plywood disc, mark the holes on the disc, and centertap the set screw locations on the shaft. Slide the flange and nipple out of the way, and drill ⅛" deep holes in the shaft where the set screws will touch. These holes should be larger than the set screw. Drill the 4 holes for the flange. (See Figure 10-7 for shaft positioning and flange side of discs.)

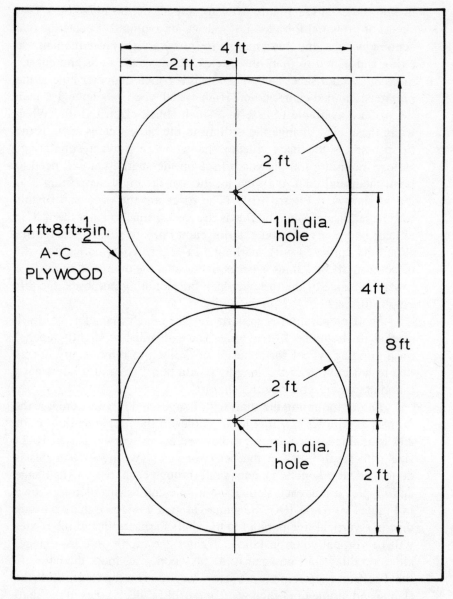

FIGURE 10-4. PLYWOOD LAYOUT

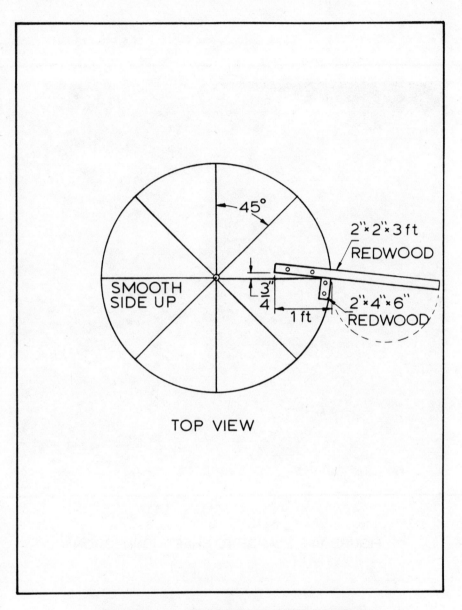

FIGURE 10-5. BLADE SUPPORTS LAYOUT

FIGURE 10-6. FLANGE TO SHAFT CONNECTION

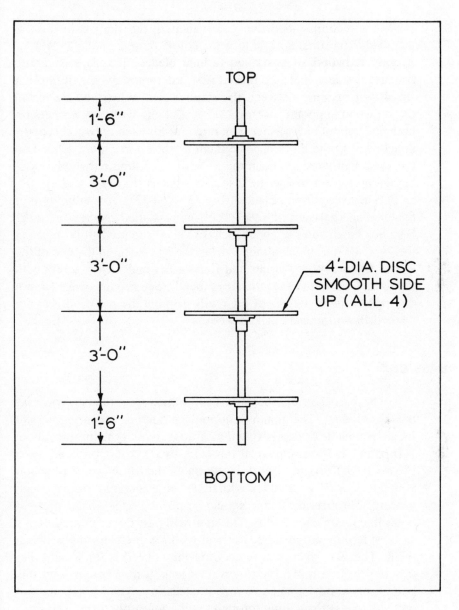

TOP

1'-6"

3'-0"

3'-0"

4'-DIA. DISC
SMOOTH SIDE
UP (ALL 4)

3'-0"

1'-6"

BOTTOM

FIGURE 10-7. SHAFT POSITIONING

It is recommended that you wait until the shaft with discs is installed in the final upright location before putting on the 2″ × 2″ × 3′ long redwood supports and turbine blades. This is a necessity because if stator walls are placed first, the turbine won't fit through the 4-foot opening between the stator walls. When you do install the redwood supports, use two ¼″ × 3″ bolts with flat washers on both ends and a lockwasher under the nut for each redwood support attachment to the discs. Locate them as shown in Figure 10-5. Use the same hardware to attach the 2″ × 4″ × 6″ long redwood blocks being careful not to use the 2 inches adjacent to the 2 × 2.

When the turbine blades (9′ × 3′ × .032) sheet aluminum or heater duct (galvanized sheet steel) are installed use a $^5/_{16}$″ × 2″ long hex head wood screw at each disc level and predrill ¼″ holes in the 2″ × 4″ × 6″ block about 1″ from the 2 × 2 and in the end of the 2 ×2. Use a fender washer and flat washer under each screw with the fender washer next to the sheet metal. (See Figure 10-8.) Do not overtighten these screws or they will strip out the redwood. Failure to predrill will cause cracked wood.

BASE BOX

The base box can be built from almost any lumber you may have available. The major requirement here is a bearing mount location capable of supporting the entire weight of the turbine (about 100 pounds). I accomplished this with two 2 × 6's on edge. (See Figure 10-9.) Figure 10-9 does not show the top piece of plywood which is 3′ × 3′ × ¾″ thick with a 1½″ hole centered for the 1-inch bearing. The bearing flange should be positioned so that 2 of the 4 holes line up on each 2 × 6. The shaft will pass between the 2 × 6's so be sure they are installed vertical with 1½-inch spacing between them. The two layer ¾″ plywood on the bottom is for weight distribution over a roof. Try to cover at least 3 roof beams with this bottom plywood—change its size if necessary. The lip hanging out on two sides is for bolting through to the roof or platform. The sides should be boarded up for rigidity but leave two doors or holes on the sides not at the ends of the 2 × 6's. These holes will be used for access when installing pulleys and the generator, and need only be open at the upper halves of those two sides.

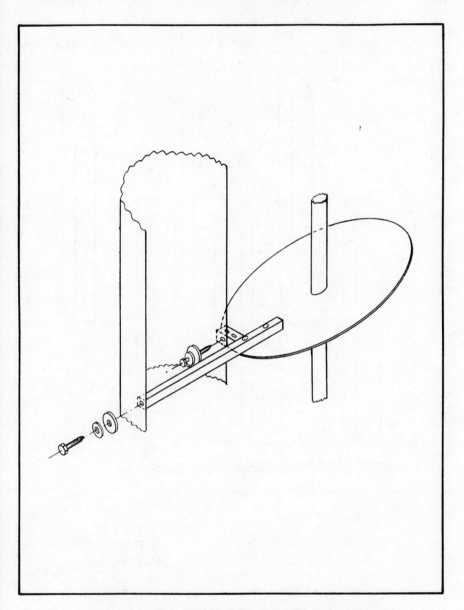

FIGURE 10-8. TURBINE BLADE ATTACHMENT

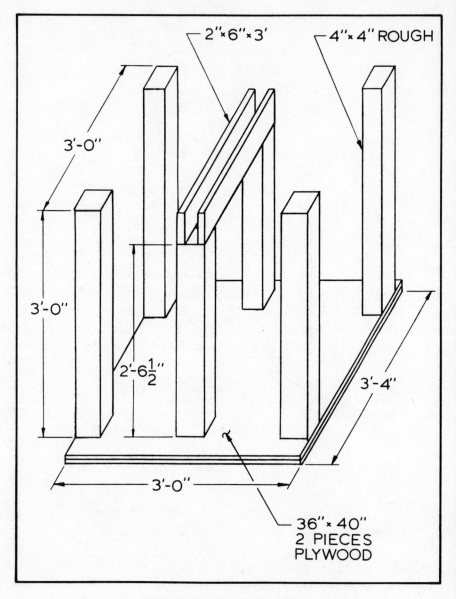

FIGURE 10-9. BASE BOX

WALLS AND TOP SUPPORT

The walls should be erected before the top support is placed, on top. (See Figure 10-10.) This particular wall set is for a 20′ × 20′ platform or roof. The walls in this case are 13½ feet long from each corner of the platform. They taper to 2 inches wide near the turbine. The walls are tangent to the 8-foot circle made by the turbine. In other words, a line perpendicular to the wall centerline should pass through the end of the wall next to the turbine and the turbine center shaft. The author's experiments show this to be the optimum location for best performance. The walls should come within 2 inches of the turbine and not touch the turbine at any time. Simple, 2 × 4 stud walls may be substituted for pointed shaped walls if desired, but slight inefficiency occurs when the wind is parallel to any one wall. The walls are 12 ft. high so as to rest on the platform and reach to the top of the turbine. The walls must be at least 8 ft. long to accomplish turbine improvement. The longer the walls, the more power, and the lower the windspeed to which the turbine will respond. With no walls the turbine will begin putting out power at 7 MPH winds. With 8 ft. walls it begins power out at 5 MPH and with longer walls lower wind response. (Note: Longer walls not tested except in scaled model. The scaled model showed that 13½ foot walls will improve power at 10 MPH winds by approximately 6 times.)

The top support is accomplished by spanning 2 × 6's between the four walls as shown in Figure 10-10. The 9½-foot lengths should be obtained from ten-foot pieces cut after installation to match the walls. One cross-member should be spaced upward at each end of the wall by 1½″ (a 2 × 4 spacer). Both cross-members should be positioned height-wise to clear the turbine. The hole through both 2 × 6's should be 1½ inches in diameter for the vertical 1-inch shaft. Locate this hole using a plumb bob to align with the bearing on the top of the base box after the base box has been centered exactly between the close ends of the four walls. Hold up on final nailing of the top support 2 × 6's until you are sure they cross at the point the hole for the shaft must be drilled, and until after the shaft and bearing are installed. Allow room at the bottom of the top supports for a 1-inch bearing or the bearing can be put on

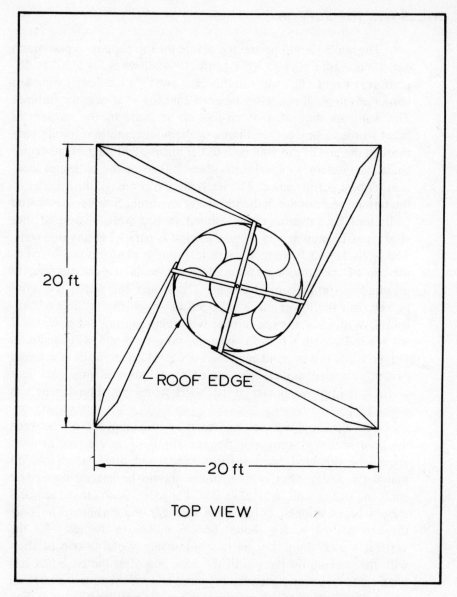

20 ft

20 ft

ROOF EDGE

TOP VIEW

FIGURE 10-10. WALLS AND TOP SUPPORT

top. When bolting the top bearing on, use four ⅜″ × 4½″ bolts clear through both 2 × 6's and provide a flat washer, lock washer, and nut on the wood end. (See Figure 10-11 for turbine mounting.) Balancing of the turbine is not required because the RPM is low (10 to 40 RPM) and the vertical shaft prevents gravitational effects from being a factor to consider.

After the turbine is installed and the blades put on, you will want to complete the funneling system by roofing over the walls. Cover everything except the 8-foot diameter hole to be left over the turbine for air escape. You will also want ramp deflectors from the platform to the bottom of the turbine. (See Figure 10-12.) These go on all four openings. All of these walls, roofs, and ramps are done in 2 × 4 framing with plywood cover (or a suitable substitute). The turbine will operate without the walls if cost delays occur, but you must then provide another method of top support for the shaft.

RPM RATIO EQUIPMENT

Because this turbine turns so slow, direct drive is not possible for generators. An RPM increase of between 10 and 60 is necessary depending on what generator you use. Basically, there are three types of RPM ratio equipment: chain and sprocket; gears; or belts and pulleys. I tried chain and sprocket first because of the no-slip drive capability. At a windspeed of 25 MPH, the lubed chain transferred vast noise levels to the turbine blades which amplified them in all directions. It sounded just like a four-engine propeller-driven bomber with the engines wide open about 50 feet over the house. Needless to say, I had to disconnect it before the neighbors complained. Because of the noise, I don't recommend this approach. Gearing was not tried because some means of lubrication is required. Instead I went to pulleys and belts which are quiet and available. I recommend about 32 to 1 RPM ratios for generators rated at 1800 RPM. If you should use a 600-RPM generator, try an RPM ratio of 15 to 1. With belts and pulleys, I found a ½-inch wide belt quite sturdy and available as an automotive accessory. For pulleys, I used ½-inch wide (for A type belts) and between 3 inches and 12 inches in diameter. Less than 3 inch diameter pulleys are available but they slip too much. I couldn't find any over 12 inches in diameter and if I could have they would be too expensive. Thus

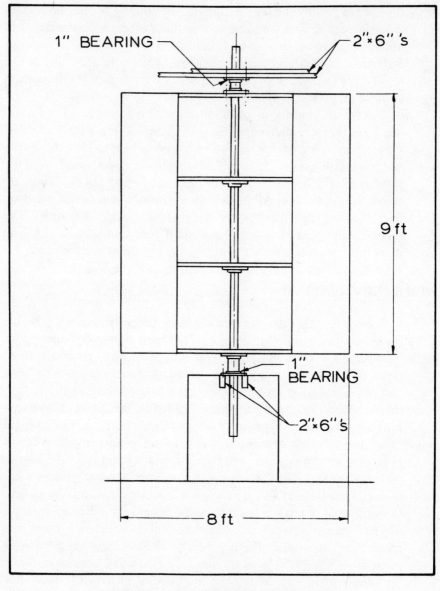

FIGURE 10-11. TURBINE MOUNTING

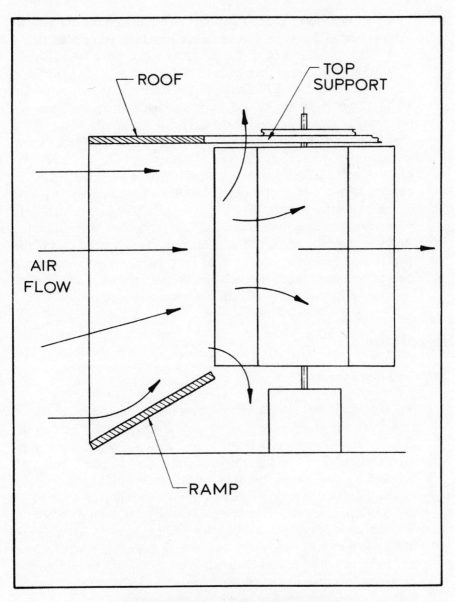

FIGURE 10-12. ROOF AND RAMPS

12 inches to 3 inches (4 to 1) was the best I could get out of one belt. With two belts, I got $4 \times 4 = 16$. With three belts you can get $4 \times 4 \times 4 = 64$ to 1 RPM ratio. I used 32 to 1 to drive a 0-to-90-volt, 1750 RPM, permanent magnet field, DC motor as a generator for a 12-volt system. This RPM ratio was obtained as shown in Figure 10-13. Two fixed shafts and two adjustable shaft positions are necessary as shown in Figure 10-14. The turbine shaft and the motor shaft are the fixed shafts. The other two are adjustable to allow loosening for replacement of belts or tightening belts. Each adjustable shaft is ½″ diameter mounted on two ½-inch bearings located on either side of a 1-inch plywood sheet. The plywood sheet is bolted or nailed to the bottom of the 2×6 pair with a 1½-inch hole provided for turbine shaft clearance. The bearings for the adjustable shafts are bolted each pair to each other through slotted holes in the 1″ plywood. Make the slots 2 inches long and in a direction that will tighten both belts simultaneously. Figure 10-5 shows a side view of how two different levels of belts exist for clearance purposes.

GENERATOR

Selecting the Generator

The generator selection is the most important item in the whole system. I am going to give you a short history of what I went through in selecting a generator. In this way, you will benefit from my experiences. I first had the idea that a car alternator would be suitable because of its low cost, ready availability, and 55 amp output. I found that it won't start putting out current until 1500 RPM and 55 amps are obtained between 4,000 and 6,000 RPM. Working at a gear ratio of 60 to 1, I found that I had to have 25 RPM on the turbine to get some power out. This occurs at above 15 MPH winds and I was unwilling to add a 4th belt to get there. So I bought a more expensive alternator that would cut in at 800 RPM. That cost me $90 as opposed to the $40 one I had bought previously.

The new, lower RPM alternator was a special unit for trucks that would put out power while in idle. I could now get power above 7 MPH winds at above 13 RPM turbine speed. I had placed a

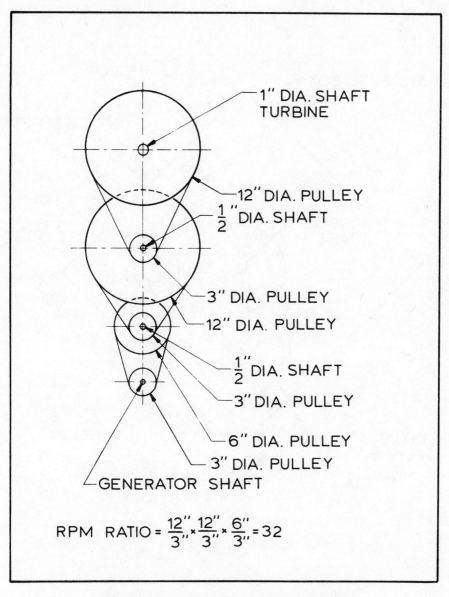

1″ DIA. SHAFT
TURBINE

12″ DIA. PULLEY

$\frac{1}{2}$″ DIA. SHAFT

3″ DIA. PULLEY

12″ DIA. PULLEY

$\frac{1}{2}$″ DIA. SHAFT

3″ DIA. PULLEY

6″ DIA. PULLEY

3″ DIA. PULLEY

GENERATOR SHAFT

$$\text{RPM RATIO} = \frac{12''}{3''} \times \frac{12''}{3''} \times \frac{6''}{3''} = 32$$

FIGURE 10-13. RPM RATIO CALCULATION

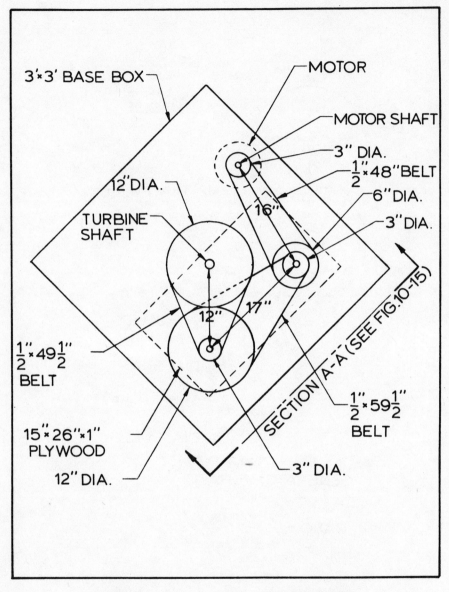

FIGURE 10-14. PULLEY ARRANGEMENT

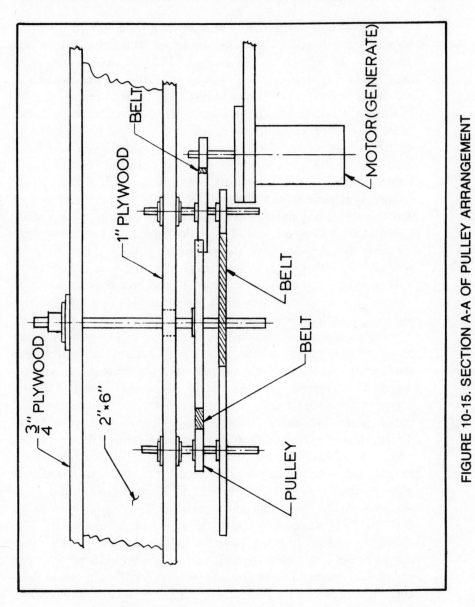

FIGURE 10-15. SECTION A-A OF PULLEY ARRANGEMENT

voltage regulator in the field circuit to limit field current when the battery was charged to 14 volts. However, when the battery was down and the RPM was less than 13 RPM, the field continued to draw 3 amps from the battery. I put a manually operated switch in the field circuit so that this could be turned off thus replacing what in an auto would be the ignition key. After that, whenever the wind soared over 7 MPH on my wind indicator, I would run to the field switch and throw it on—charging the batteries. I even asked my wife to do this while I was at work. Soon I noticed that this was a bother, and if forgotten when the wind died, would discharge the battery more than it been charged. Obviously, an automatic circuit was in order. I dug through my junk box and came up with a relay and a small DC permanent magnet field motor. I belted the little motor to the alternator and ran it through a diode to the relay. It worked out so that the relay was actuated by the little motor acting as a generator and the relay contacts turned on the alternator field current only when sufficient RPM was available. That chore out of the way, I settled back to see the results.

I soon found that, although I only turned on the alternator when it was going to put out current, it very often put out less current than the 3 amps that the field was drawing from the battery. After some careful data collection, I found that the winds must be at 10 MPH or higher before I would exceed 3 amps output. I could adjust the field cut-in point and did so, leaving my system with a 10 MPH threshold. In my mind this was totally unacceptable. I decided to get a larger PM DC motor and drive the alternator field with it so as not to drain the battery. I did this, but found that now with less than 3 amps on the field in winds below 10 MPH, the output of the alternator was so low that it failed to come up to 13-volt charging voltage until 10 MPH was reached! Obviously, I needed a bigger DC PM motor. I was still throwing away the first 36 watts of power in the alternator field, showing low efficiency at low RPM. Since a bigger permanent magnet DC motor was needed for the alternator field current, I decided that I might as well use it to drive the battery, charging it directly. So I ordered a ¼-HP PM DC motor rated at 90 VDC, 1750 RPM, 3 amp armature current. I wanted one rated at 24 VDC, 600 RPM, 30 amp armature current, but couldn't stand the 6-month lead time from the factory.

With the 1750 RPM motor on hand, I dropped back to an RPM ratio of 32 to 1. Because of the 90-volt rating, the motor turning as a generator would start putting our current at a turbine speed of 10 RPM (13 volts), which is 7 MPH winds and would reach its maximum current rating of 3 amps at 10 MPH. Well, that solved the 36-watt loss below 10 MPH winds. But, if you noticed, I have *another* problem! At 10 MPH winds, I reach my maximum, continuous armature current. Actually I can go to 6 amps intermittantly with no motor damage so I put a 6-amp slow-blow fuse in series with the motor output. Since I am behind a hill I seldom see winds over 10 MPH, so little fuse blowing occurs.

I have used this particular test facility in obtaining power at wind speeds below 7 MPH, by adding the walls mentioned earlier in this chapter. The $90, DC, 3 amp motor is not a loss, however, as it turned out to be the ideal generator for my water wheel. For your turbine, try to get a permanent magnet field, DC motor rated at twice the voltage of your battery system, an armature current as much above 20 amps as you can find, 600 RPM, and about 1 horsepower. These can be purchased from: Applied Motors Inc., P.O. Box 106, 4801 Boeing Drive, Rockford, Illinois 61105, phone 815-397-2006. They make them for electric motorcycles, electric forklifts, electric golf carts, etc. You may also find some stocked where parts are available for the items mentioned. If you are successful at locating a 600 RPM motor you can reduce your RPM ratio to between 10 and 16 to 1. This is a two-belt system as opposed to 3 belts.

To wire a DC motor as a generator is easy. (See Figure 10-16.) The diode keeps the motor from running as a motor and allows generated current to charge the battery. This diode should be rated at twice the amps of the armature current and twice the voltage of the battery. Polarity on the DC motor is usually positive on the red wire if the shaft is turned clockwise (facing it) and negative when turned counterclockwise. To be sure which is which, put a voltmeter on it and turn the shaft clockwise. This wind turbine has all pulleys and shafts turning clockwise when viewed from above.

Any motor available in this class should run you between $100 and $125. If they are priced higher, ask for a discount. If a discount is not given, compare elsewhere. Normal list prices run around $250

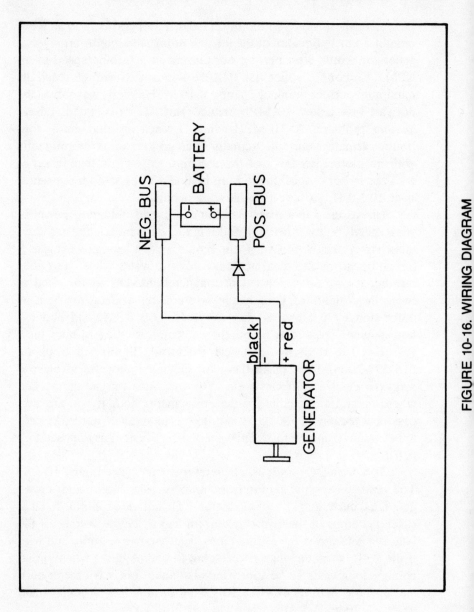

FIGURE 10-16. WIRING DIAGRAM

and I know of no one that pays list price. You may get a larger discount when purchased through the company you work for—if they will do it for you.

In review, this turbine is designed for roof-top or hill-top applications. With long enough walls you may obtain power below 5 MPH winds. The PM, DC motor output with diode is completely automatic for charging batteries; no current flows until charging voltage is reached. Furthermore, this type of turbine is self limiting in RPM during high winds. Mine stabilized somewhere around 40 RPM in 30 MPH winds without vector walls. With vectoring walls, **rotational speeds over 60 RPM are not expected. Therefore, 60** RPM times the RPM ratio should equal your generator RPM rating if you have long vector walls. You should use higher RPM ratios for short walls or no walls.

I have performed $1,000 worth of experimenting in bringing this vertical wind turbine to this point so that you would not have any expensive mistakes in your system. You should be able to build it for between $300 and $500 depending on the extent of your walls and an available platform.

SAFETY

This sytem is relatively safe provided you heed certain precautions *at all times*. Do not put your hands near operating pulleys. Provide a lock or brake for the turbine, so that you can safely work in the pully and generator area; do not rely on the wind remaining calm during maintenance. The principle dangers are shorts and body contact with moving parts. Shorts at the generator connections will not draw battery power if your in-line diode is in place, but generator power will be shorted if operating. Take special precautions to prevent the positive and negative leads from coming together. Results of this could be a burned out armature winding and possible arc burns to you if your hands are close. The turbine turns such that the rounded part of the blades will strike you if you walk into it. Although this is not as hazardous as a sharp edge, damage to the turbine is more likely than to yourself. The turbine access should be gated to prevent children from approaching the moving turbine.

My approach to maintenance when the turbine is moving has been to stop the turbine with my hands by bouncing off the turbine blades slowing it a little at a time until I could grab it and insert two long bolts between the turbine and base. I once released it in winds 15 MPH with gusts to 25 MPH and as I withdrew the locking bolts, I hung onto a lower arm supporting a turbine blade. I found that in winds over 15 MPH, one person cannot hold the turbine against the wind (I was dragged around the floor slowly until I let go). Although I was not hurt or even scared, I was greatly surprised at the amount of torque created. If you have to hold the turbine over 15 MPH winds, use two or more people. It takes 4 people in 25 MPH winds! Treat this torque with great respect!

11.

Building the Landing Propeller-Type Wind Power Plant

DETAILED DESCRIPTION

This propeller wind system is a result of Bob Landing's research into a practical propeller windmill that you can make yourself. Bob lives in Pleasant Hill, California and deserves full credit for the information on how to build this particular wind system. The completed unit is shown in Figure 11-1. It has an 11½-foot diameter, 3-bladed wooden propeller. The propeller drives a spur gear with an RPM increase of 7½ to 1 driving a PM DC motor. Output can exceed 40 amps when loaded at 15 volts DC (600 watts). The propeller is guided into the wind by an automatic folding tail with gravity return. The tail folds at about 35 MPH winds to protect the propeller from excessive speeds. The tail can also be folded manually with a cable from the ground. The entire system is mounted atop a ground-mounted, 45-foot mast. This system operates in winds of 7 to 35 MPH.

The fixed-pitch propeller blades are made of vertical-grained, Sitka spruce and painted with 3 coats of asphalt aluminum paint. Good quality pine can be substituted. Each blade is 5 ft-10 inches long and is made from two 1×6 boards glued together. This propeller will drive loads of 1800 watts in 20 MPH winds using a larger generator (the author's calculation).

The gearing is direct spur-gear drive to the generator. One 10-inch diameter 100-tooth spur gear mounts on the propeller shaft and drives a 14-tooth smaller gear that is on the generator shaft. These gears run in a homemade housing enclosing an oil bath. This

175

FIGURE 11-1. PROPELLER WIND SYSTEM

housing (gear box) is made of surplus aluminum ¼ inch thick with spacers for gear box bolts. (See Figure 11-2 and 11-3.) Two plates are required to make this gear box. To keep in the oil and reduce noise, wooden strips were bolted tightly on all four sides to retain oil. Other methods may be used to retain lubrication if desired.

The mast support is a 1-inch pipe inside of a 1¼-inch pipe. These act as a bearing assembly and vertical support for wind direction changes. The collector rings are brass from a brass pipe. The collector brushes are from an automotive starter. Two are used on each ring. Almost any hard carbon brush will do. The brushes with their holder are attached to a 1-inch angle iron and the entire assembly rotates around the collector rings.

The tail unit is off-center from the mast support and is hinged with a 10-degree tilt so that it will fold toward the side automatically in high winds (35 MPH and higher) or manually when you wish to stop the propeller rotation for maintenance or power shutdown. The tail fin is $^3/_{32}$-inch thick × 17″ × 32″ mounted on a ¾-inch pipe slid into a 1-inch pipe.

The tower used was a retired ham radio antenna mast 45-feet tall. As mentioned in an earlier chapter, it is best if your wind turbine is 15 feet above all obstacles and trees within 400 feet. The tower should be sturdy (a triangular shape is recommended). Support should be on the ground, on a concrete pad (to prevent noise transfer to the house). Guy wires should be used with heavy turn buckles and heavy wires.

This system will charge batteries of 6-volts or 12-volts. The 6-volt battery can be charged at lower wind speeds since a diode is used in the output. When the generator voltage exceeds the battery voltage by about 0.6 volt (diode voltage drop) current begins to flow. Prior to current flowing, the generator expends its energy in increasing the voltage only. Therefore, lower voltages are reached first and current flows earlier than for higher voltages. (See Figure 11-4 for a parts list of the entire wind power plant.)

PROPELLER INSTRUCTIONS

Step-by-Step Building Instructions

A 2-blade prop is good for direct drive up to 9 feet in diameter.

(Photo courtesy Bob Landing)

FIGURE 11-2. EXPOSED GEARS

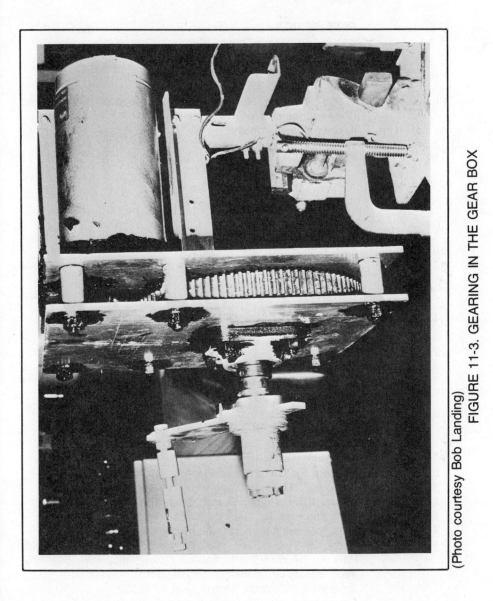

(Photo courtesy Bob Landing)

FIGURE 11-3. GEARING IN THE GEAR BOX

QTY   **DESCRIPTION**

PROPELLER PARTS

6	1″ × 6″ × 6′ long, Sitka spruce, no knots and ck. grain vertical. (Pine second choice)
1 gal.	Asphalt-aluminum paint
1 qt.	Weidwood dry powder glue
3	Copper, brass, or aluminum shim stock, 6″ × 36″

PROP HUB PARTS

1	Aluminum plate, ¼″ × 13¾″ × 15¾″
3	Aluminum plates, $3/32$″ × 5½″ × 6″
15	Bolts, $5/16$″ dia. × 4″ long or less to fit, with flat washers, lock washer, and nut.
1	Shaft, solid steel, 1″ dia. × 11½″ long, threaded one end, stainless steel preferred.
1	Hub, for 1″ shaft to prop hub
1	Bolt ¼″ dia. × 1″ long
2	Flat washers, 1″
1	1¼″ pipe spacer × 2″ long
1	Castle nut, 1″ with cotter pin (from auto wheel axle)

GEAR BOX, SUPPORT SHAFT, AND COLLECTOR RING PARTS

2	Aluminum plates, ¼″ × 11″ × 14″
6	Spacers, ½″ dia. aluminum pipe × 1¾″ long
6	Bolts, $7/16$″ dia. × 3½″ long with nuts and lock washers
2	Bearings, for 1″ shaft, Boston Gear 0694B-5F
8	Bolts and nuts, for bearings (or drill and tap for bolts)
4	Bolts for mounting motor with nuts and washer
1	Motor, 36 VDC, ¾ HP, 1800 RPM w/permanent magnet field, Part no. BA3637-318-1. Applied Motors, Inc., P.O. Box 106, 4801 Boeing Dr., Rockford, Illinois 61105.
1	Spur gear, 100 tooth, Boston Gear NF100
1	Spur gear, 14 tooth, Boston Gear NF148
4	Wood strips, to fit, rubber paint coated

FIGURE 11-4. PARTS LIST

QTY	DESCRIPTION
1 cup	90 W gear oil
1	Angle iron, 2″ × 2″ × 14½″ long
2	Angle iron, 2″ × 2″ × 10¾″ long
4	Bolts, lock washers, nuts, ⁵/₁₆″ dia. × 1″ long
1	Pipe, galv. steel, 1″ dia. × 27″ long
1	Pipe, galv. steel, 1¼″ dia. × 15″ long
2	Bolts, lock washers, nuts, ⁵/₁₆″ dia. × 3″ long
1	Aluminum plate, ¼″ × 8″ × 13¼″ and mount hardware
4	Brushes and holders (from auto starters)
2	Brush holder brackets (copper)
1	Fiberglass board (for brush brackets)
2	Angle iron, 1″ × 1″ × 4″ long
1	Pipe, galv. steel, 1¼″ dia. × 4′ long
1	Terminal block, 2 point
1	PVC pipe, 1½″ dia. × 6″ long (insulator)
2	Brass pipe, 2″ dia. × 1″ long (collector rings)
10 ft.	Wire, #10
1	Pipe flange 1¼″

TAIL PARTS

QTY	DESCRIPTION
1	Aluminum plate, ¾″ × 7¾″ × 10″
1 pair	Door hinges, heavy duty steel with machine screws or bolts
1	Aluminum sheet, ³/₃₂″ × 17″ × 32″
1	Pipe, galv. steel, ¾″ dia. × 6½′ long
1	Pipe, galv. steel, 1″ dia. × 2′ long
1	Pipe coupling, 1″
2	Aluminum straps, 1″ × ¼″ × 18″
1	Aluminum plate, ¼″ × 3″ × 13″
1	Bolt, ¼″ dia. × 3″ long with lock washer and nut
1	Tail stop bolt, ⁵/₁₆″ dia. × 4″ long with nuts
2	Bolts, ¼″ dia. × 2½″ long
2	Angle iron, 1″ × 1″ × 6″
2	Pulleys, small (for tail cable)
60 ft.	Plastic coated steel cable (for tail folding)
1	swivel, (for cable)
1	2-pound weight (for cable)
1	Hook, (for cable)

FIGURE 11-4. Continued

QTY   **DESCRIPTION**

TOWER PARTS

4 (variable)	10-foot, 3-legged tower section, Rohn 20G
1	9-foot, 3-legged, tower top section Rohn 20AG
1	Base plate, Rohn BPC25G
6	Turnbuckles
300 ft.	Guy wire, heavy
60 ft.	Wire, two conductor, #6, insulated for outdoor use

PANEL PARTS

1	Diode, 50A, 50V, 1N248
1	Heat sink for diode
1	12-volt light and socket
1	Voltmeter, 0-20 volts DC
1	Ammeter, 0-50 amps DC
1	Selector switch, for voltmeter
1	Fuse, 40 amps screw in house fuse
1	2-pole disconnect (Always be sure tail is folded before disconnecting load).

FIGURE 11-4. Continued

It is recommended that a 3-blade prop be used from 9 feet through 17 feet in diameter for direct or gear drive.

Bob Landing has provided this Prop-Speed Chart in Figure 11-5 based on a maximum tip speed of 125 MPH for given diameters. Note that a prop that reaches 125 MPH tip speed is at its maximum horsepower. Any tip speed over 125 MPH does not increase power. It is unsafe if the tip speed reaches 150 MPH (for diameters 10-feet or larger).

The prop can be made of almost any good, strong wood that is not too heavy in weight. The grain must be vertical, as in Figure 11-6A and not like Figure 11-6B which bends easily. If you can't buy vertical-grain spruce wood, then try pine or douglas fir. But the boards must not bend easily.

PROP DIAMETER	PROP RPM	NO. OF PROP BLADES	MADE FROM MEASURED BOARD SIZE OF: (End View)
6'	600 RPM	2	¾" × 3"
8'	437 RPM	2	1" × 4"
10'	349 RPM	3	1" × 4"
10.5'	334 RPM	3	1¼" × 5"
11'	318 RPM	3	1¼" × 5"
11.5'	304 RPM	3	1½" × 5½"
12'	291 RPM	3	1¾" × 5½"
12.5'	280 RPM	3	1¾" × 5½"
13'	269 RPM	3	1¾" × 6"
14'	250 RPM	3	1¾" × 6"
15'	233 RPM	3	1¾" × 6½"
16'	219 RPM	3	1¾" × 6½"
17'	206 RPM	3	2" × 8"
18'	194 RPM	4	2" × 8"
30'	117 RPM	4	Sail-type blade

FIGURE 11-5. SAFE PROP RPM FOR 125-MPH TIP SPEED

Now, let's say you are going to build a *12-foot-diameter* prop blade. The blades will each have to be 6 feet long, and three are required for one propeller. Buy either: (1) 2" × 6" × 6' board; or (2) two 1" × 6" × 6' boards for each blade (to be glued together) with no knots in boards. Be sure to pick *dry, cured* wood. This is a must; otherwise the prop may warp or split. Also, buy some cheap boards for practice; make a prop blade and find your mistakes in doing so.

Better balance in the blades can be obtained with choice 1" × 6" boards; it requires 2 of these boards for one blade. They must be glued together with Weldwood dry-powder glue. Mix with water and apply to both boards. Allow to set a little before joining them and clamping them in 12 "C" clamps overnight to dry. Glue one each night. Six boards are needed for the 3 blades. Make 4 blades just in case you spoil one. After the boards are dry, plane and square them so that all boards are the same dimensions. Also, check each one for weight on a kitchen scale. They may weigh around 12 pounds each.

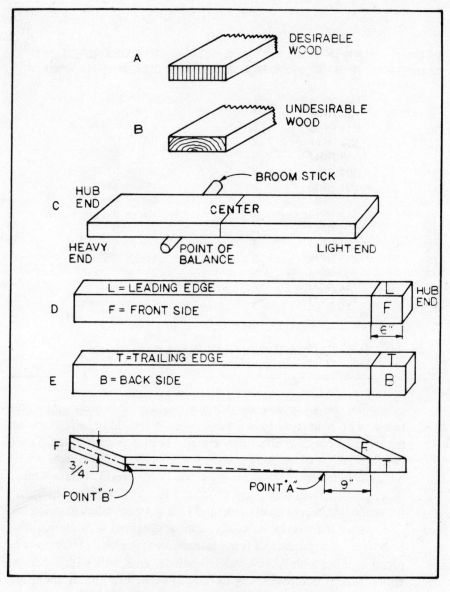

FIGURE 11-6. PROPELLER LAYOUT

Make a record of the weight of each board. When finished and ready for paint, each blade will weigh about 5½ pounds.

Now use the teeter-totter test as shown in Figure 11-6C, to determine the hub end of the boards. Get a broom, or something that is round, and find the balancing point on the board in relation to the center of the board. If it balances somewhere off center, this shows that one end is heavier than the other. The heavier end will be the hub end. Plainly mark it on the board by means of a line drawn all the way around the board, 6 inches from end. (See Figure 11-6D.)

Commercial manufacturers of wind-driven propellers make a straight-cut propeller blade because it is low in labor cost. It has proven to be a reasonably reliable propeller for general use. The secret is that the front of the blade is flat all the way from the hub to the tip, with hardly any tapering off in width, but it is slightly rounded off on the leading edge. What we start with is flat all the way, with *no work to be done on this side of the board* except for sanding.

All we have to do is work on the back side of the board, the side that faces the tail. At the hub end, mark with ink in big letters the letter F which means that this side of the board will be the front or flat side of the prop blade, and no work will be done on it. (See Figure 11-6D.) On the edge of the board mark the letter L which means the leading edge of the blade. Do this in ink also on the hub end of the board. Now turn the board over and mark the opposite side of the board with the letter B, as shown in Figure 11-6E. This letter means the back side of the blade. Now mark the other edge of the board with the letter T, which means the trailing edge of the board. This is the sharp side when the blade is finished. Do this at the hub end in ink. Turn the board edgewise, as indicated in Figure 11-6F and draw a line on the trailing edge from point A to point B. Point B will be ¾ inch down from the front of the board, and point A 9 inches from the 6-inch mark on the hub end of the board. The line at B should extend across the end of the board parallel to the front side. Cut off the bottom wedge, which is the *back of the board*. The board will look something like a small diving board after this cut.

In Figure 11-7H, mark points A and B, and draw a line connecting them. Point B is ¼ inch down from the *front* of the board on

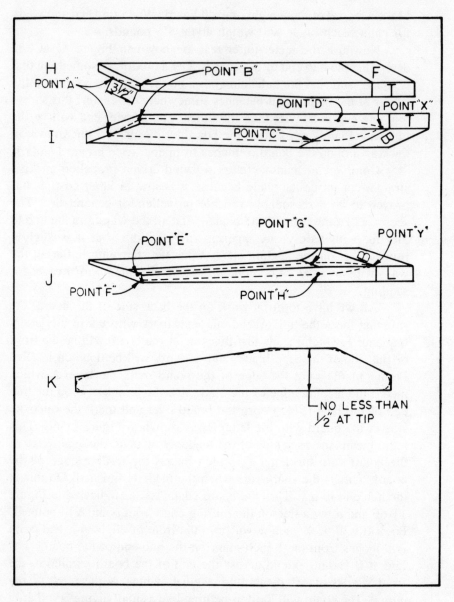

FIGURE 11-7. PROPELLER CUTTING AND SHAPING

the trailing edge and point A is 3½" inches from point B on the back of the board. This is the trailing edge you are working on.

On the back and trailing sides mark points C and D, as indicated in Figure 11-7I. Point C is 3½ inches from the trailing edge of the board and 15 inches from the hub end of the board. Draw a line between points A and C. Point D is ¾ inch from the front of the board and 15 inches from the hub end. Draw a line between points B and D. Mark point X 6½-inches from the hub end of the board as in Figure 11-7I. Draw a line as shown from point C through point X to point D. Plane and file out this area. Now saw on the straight line area until you reach points C and D.

Turn the board with the back side up as shown in Figure 11-7J. Mark these points. Points E and G are marked 1 inch from the *leading* edge on the back side of the board. Point F is ¼ inch from the *front* side of the board on the *leading* edge. Points G and H are 15 inches from the hub end of the board. Point H is ¾ inch from the *front* side of the board on the *leading* edge. Point Y is 6½ inches from the hub end. Draw a curved line as indicated through points G and Y to H. File and plane out this area. Draw lines G to E, E to F and F to H and saw off this wedge.

Figure 11-7K is a drawing of the tip shape before the corners are rounded off. File and plane the board to make it round on the back side *only*, as shown by the dotted lines. The front side still remains flat, with no work on it at all except for smoothing and sanding.

Shape the back side as indicated in Figure 11-8 L so that it is smooth and tapered off as it goes to the tip. Make all 3 prop boards exactly the same shape, and keep their weight the same as much as possible.

Figure 11-8 M shows how to make an 11-degree (15 degree for average winds over 23 MPH) wedge and mount it on the prop board. The wedge is glued on with Weldwood glue and allowed to stand overnight while clamped. Figure 11-8 N shows how to shape the tip end to complete the prop. The trailing edge is slightly tapered for about 6 inches so that the end of the blade is about 3½ inches wide and about ½ inch thick near the leading edge. Make a slightly rounded leading edge.

If you wish, you may protect the prop leading edge with copper, brass, or aluminum shim stock about 3 feet long, tacked or

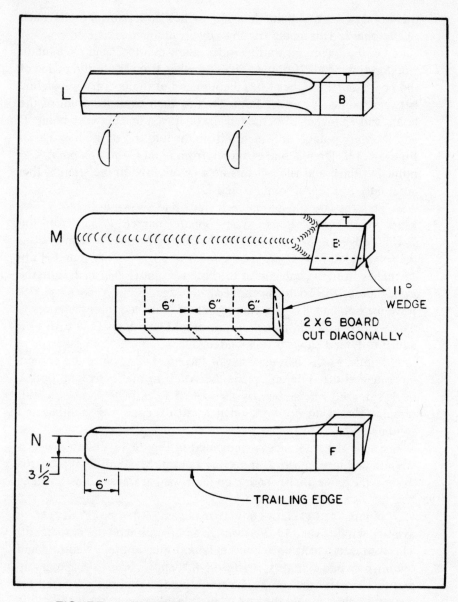

FIGURE 11-8. COMPLETING THE PROPELLER BLADES

stapled, and glued around the leading edge near the tip area. This tip area gets hot at speeds near 125 MPH. If you live in a dusty or sandy area, you need this protection on the leading edge of the blade; otherwise it will sandblast your prop to the bare wood. Do this after shaping, filing, and sanding and before painting. When smooth, weigh all 4 blades and select the 3 best blades. Paint them with *3 coats* of asphalt aluminum paint. At first, the paint appears to be black, but when it is dry, it is a bright aluminum finish. Saturate the blades with the paint. It seals the wood from moisture.

Figure 11-9 shows how to make your hub plate and mount your blades on the plate. Once the blades are mounted on the hub, mark each blade with a number for identification to match its resting place on the hub.

To balance the blades, use a bubble balancer such as is used in tire shops or in gas stations to balance wheels; and as you balance the prop, add necessary weights on the first 2 bolts nearest the tip of the blade, using thick steel-strap stock drilled with holes to fit the bolts holding the prop of the hub plate. The weights can be on the back side of the hub plate for a neater appearance. You will have to experiment with different weights and see what is needed. Do this where there is no breeze.

THE PROP HUB

Use not less than ¼″ thick aluminum stock material, 13¾″ × 15¾″ for the hub plate. (See Figure 11-9.) Paint all steel outside parts with asphalt aluminum paint. Do not cut out the area between prop blades on the hub plate as it could result in failure under high wind loads. Each blade will have an 11-degree wood wedge glued under it to tilt the flat front when mounted. Be sure the flat side of the blade faces out toward the wind. The hub plate is between the blades and the gear box, and the individual prop blade plates (as shown in the corner of Figure 11-9) are mounted on top of each blade as shown in Figure 11-10. Use five ⁵/₁₆″ bolts with flat washers at both ends and a lockwasher on the nut end for each prop blade. Bolt length will vary, select to fit. After the propeller assembly is bolted together, balanced, and painted it can be mounted to

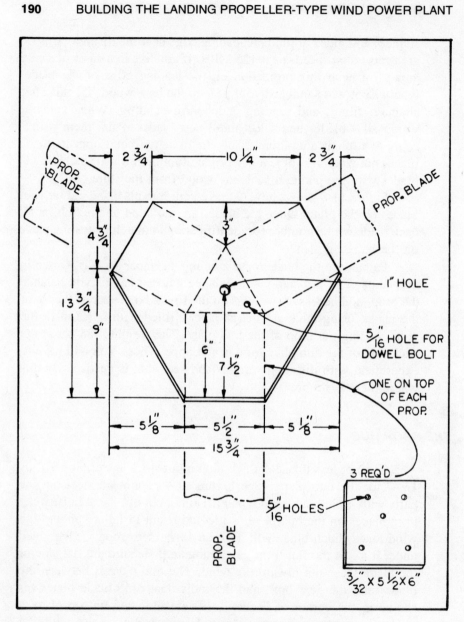

FIGURE 11-9. PROPELLER HUB

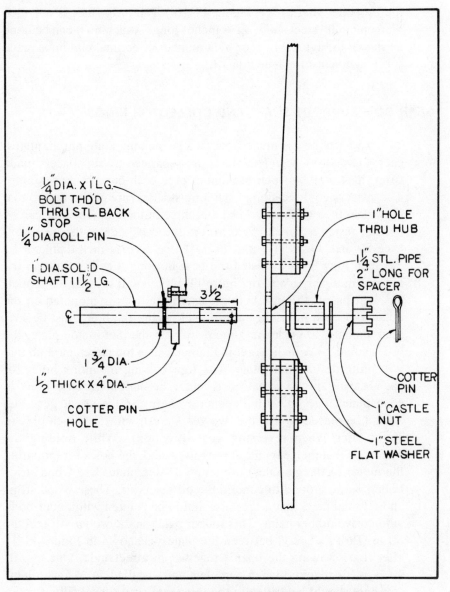

FIGURE 11-10. PROPELLER ASSEMBLY MOUNTING

the propeller shaft. (See Figure 11-10.) The prop shaft is a 1-inch diameter solid steel shaft, 11½ inches long. Attaching it can be done as shown in Figure 11-10 or by a number of commercial hubs made for attachment to a 1-inch shaft.

GEAR BOX, SUPPORT SHAFT AND COLLECTOR RINGS

The gear box is made from two ¼-inch thick aluminum plates, each 11 × 14 inches. The plates are separated by six spacers made from thick wall, ½-inch aluminum pipe each cut 1¾ inches long. The spacers go three along each long side and are positioned so that they clear the two gears and so that three of them on one side can be used to fasten a 2-inch angle iron to the back side of the gear box. (See Figures 11-2, 11-3 and 11-11.) Use six $7/16$-inch diameter by 3½ inch long bolts to fasten the gear housing together through the aluminum pipe spacers. In Figure 11-11 you will see that the propeller shaft passes through both plates with a bearing mounted on the outside of each plate. In the prop-side plate, the hole must be 1¼-inch diameter to clear the 1-inch shaft. In the motor-side plate, the hole should be ⅛ inch larger in diameter than the large gear hub that will protrude into that plate. The four bearing mounting holes for each bearing should be placed to center the shaft in the large holes in each plate. The hole for the motor shaft and the small gear hub should be placed so that the two gears mesh with .010 inch play (a slight click when reversing gear direction). While holding the motor, so that the gears mesh properly, mark the holes for mounting the motor (4 flange holes) to the motor-side plate. Use wood strips bolted on all around to contain the oil reservoir. These wood strips and all steel parts that might rust should be painted with under-body automotive rubber paint. This rubber paint on the wood will seal the oil in. Bolt the wood between the plates as shown in Figure 11-11 after first screwing the boards together in a rectangle. One cup of 90 W oil is required when complete and installed. A ½-inch diameter hole should be drilled in the top wood strip for oil filling and a tapered wood plug put in as a stopper plug. Oil is added only after a two-to-three-day dry run to break-in the gear mesh.

To mount the gear box to the support shaft see Figure 11-12. Cut and attach a 2-inch by 2-inch angle iron, 14½ inches long to the

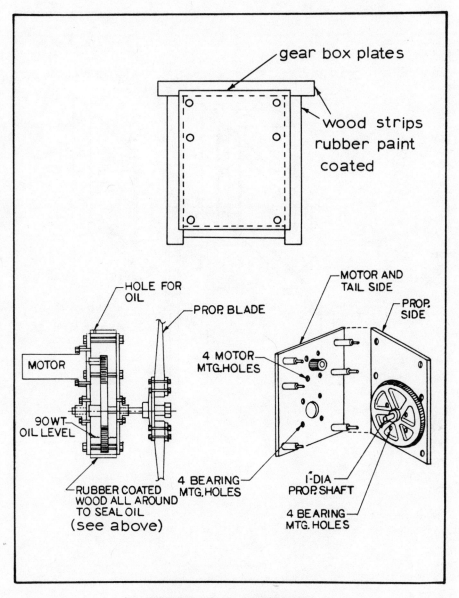

gear box plates

wood strips
rubber paint
coated

HOLE FOR
OIL

PROP. BLADE

MOTOR

90 WT.
OIL LEVEL

RUBBER COATED
WOOD ALL AROUND
TO SEAL OIL
(see above)

MOTOR AND
TAIL SIDE

PROP.
SIDE

4 MOTOR
MTG. HOLES

4 BEARING
MTG. HOLES

1"-DIA.
PROP. SHAFT

4 BEARING
MTG. HOLES

FIGURE 11-11. GEAR BOX

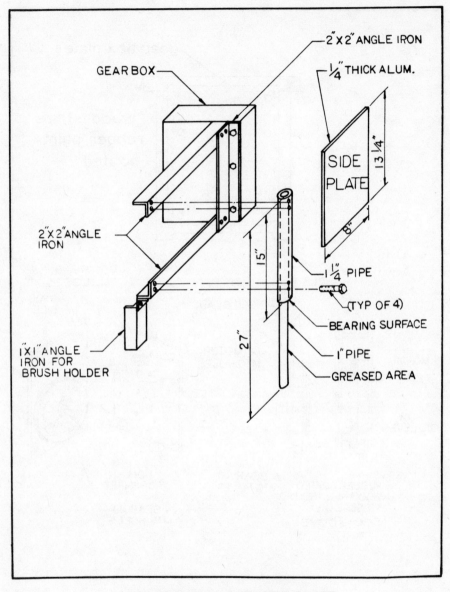

FIGURE 11-12. SUPPORT SHAFT

right rear side of the gear box using three of the $^7/_{16}$ inch bolts existing through the box. To this angle iron attach two $2'' \times 2''$ angle irons, 10¾ inches long extending rearward as shown in Figure 11-12. Use two $^5/_{16}$-inch diameter $\times$ 1 inch long bolts at each attachment point. Mount the support shaft to the ends of the two extending 10¾-inch long angle irons as shown in Figure 11-12. The support shaft is made from a 27-inch long piece of 1-inch (inside diameter) steel galvanized pipe inside of a 15-inch long, 1¼-inch diameter pipe. The two pipes are mounted one inside the other flush at the top as shown in Figure 11-12. Bolt through both pipes and the angle iron using two $^5/_{16}$-inch diameter by 3-inch long bolts at each angle iron. The bottom cut on the 1¼-inch pipe must be both perpendicular to the pipes and smooth as this is the bearing surface for rotation of the windmill into the wind. The side plate $13¼'' \times 8'' \times$ ¼" mounts inside the angle irons for rigidity.

The brushes are obtained from salvaged automobile starters. You need a total of four brushes, two for each collector ring. The brushes are mounted on brush holders as shown in Figure 11-13. Two brush holders are mounted on a U-shaped bracket made of copper or brass stock for soldering. The two U-shaped brackets, one for each collector ring, are mounted on a fiberglass board (or other insulating material). The fiberglass board is mounted to brackets that are attached to the support shaft above in any convenient fashion so that the brushes turn with the gear box and its bracketry. The brushes should be set so that they rub smoothly on the collector rings with the holders clearing moving parts by $^1/_{16}$ inch. Each brush must have a flexible copper wire coming out of its top to bolt or solder to the copper holder and a spring to hold the brush against the collector ring. (See Figure 11-13.)

The lower fixed support shaft is 1¼-inch diameter pipe of sufficient length to mount to the tower and to have a foot or more of pipe above its tower for collector rings and wiring terminal points. Use a piece of PVC (polyvinyl chloride plastic) pipe cut lengthwise down one side for an insulator over the 1¼-inch support pipe and under the brass collector rings. The collector rings can be made from a brass pipe using about ¾-inch wide, cut off pieces. Solder two #8 or #10 wires to the edge or inside of each ring and bring the two wires down thru the PVC pipe slit to the two terminal posts (insulated terminal strip). (See Figure 11-13.)

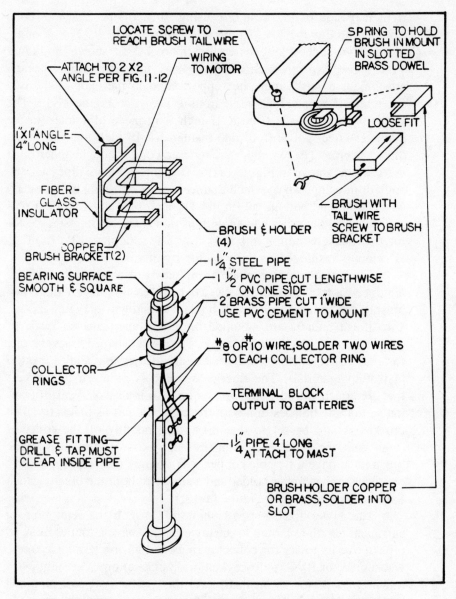

LOCATE SCREW TO REACH BRUSH TAIL WIRE

WIRING TO MOTOR

ATTACH TO 2 X 2 ANGLE PER FIG. 11-12

SPRING TO HOLD BRUSH IN MOUNT IN SLOTTED BRASS DOWEL

1"X1"ANGLE 4" LONG

FIBER-GLASS INSULATOR

LOOSE FIT

BRUSH & HOLDER (4)

BRUSH WITH TAIL WIRE SCREW TO BRUSH BRACKET

COPPER BRUSH BRACKET (2)

BEARING SURFACE SMOOTH & SQUARE

$1\frac{1}{4}$" STEEL PIPE

$1\frac{1}{2}$" PVC PIPE, CUT LENGTHWISE ON ONE SIDE

2" BRASS PIPE CUT 1" WIDE USE PVC CEMENT TO MOUNT

#8 OR #10 WIRE, SOLDER TWO WIRES TO EACH COLLECTOR RING

COLLECTOR RINGS

TERMINAL BLOCK OUTPUT TO BATTERIES

GREASE FITTING DRILL & TAP MUST CLEAR INSIDE PIPE

$1\frac{1}{4}$" PIPE 4' LONG ATTACH TO MAST

BRUSH HOLDER COPPER OR BRASS, SOLDER INTO SLOT

FIGURE 11-13. BRUSHES AND COLLECTOR RINGS

The tail assembly is constructed as shown in Figure 11-14. First, make the tail plate from ¾" thick aluminum (7¾" × 10" starter) and mount it to the mast as shown in Figure 11-14. The offset from the shaft center is important. Now construct the tail assembly using steel door hinges, aluminum, and pipe as described in Figure 11-14. The tail fin is made from $3/32$ inch thick by 17" × 32" aluminum plate. After it is installed on the pipe, the adjustable slip of the ¾" pipe inside the 1" pipe should be adjusted so that the overall length from the end of the 1-inch pipe to the tip of the tail fin is approximately 8 feet, 4 inches. This should nearly balance the propeller. Bolt the two pipes together after the adjustment is made. Continue tail construction as in Figure 11-14.

The cable for manually folding the tail is attached to the tail arm about a foot out from the hinges, brought through two pulleys and a slot to the inside of the vertical support shaft, and hence down the center of the tower with swivel joint, 2 pound weight on the end, and a hook. The tail stop that hits the mast must be greased to prevent squeaking. Fasten a cut off glove finger filled with grease over it.

The tower is a 45-foot high Rohn 3-legged tower. Set it on a concrete slab and guy wire it using turnbuckles. Failure to run at least 3 guy wires to each of two levels could result in the tower falling in high winds. Don't take any short cuts on anchoring your tower—it will pay off in the long run. (See Figure 11-15 for installation and 11-4 for parts.)

The electrical hookup for the windmill is simple. (See Figure 11-16.) A diode as shown in the parts list is used in series with the batteries you are charging to prevent the batteries from driving the motor. Optional indicator light, ammeter, and voltmeter are shown to measure results. Use at least two batteries with this unit to prevent boiling dry.

This concludes the building instructions on this unit. You may see better or easier ways to build this system. Understand what you are accomplishing and make changes as you wish. One critical area is the folding tail. It is designed to fold at a certain wind speed. Essentially, wind loading on the propeller causes it to try to turn counterclockwise as viewed from above. (See Figure 11-17.) When this happens, the tail stays aligned with the wind, but must hinge upward 10 degrees. Therefore, the prop thrust is overcoming the

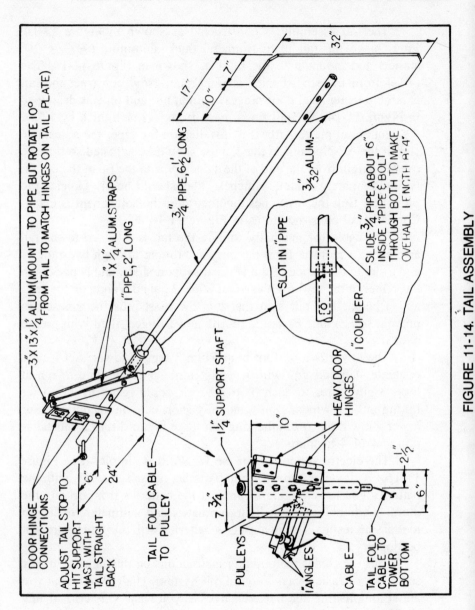

FIGURE 11-14. TAIL ASSEMBLY

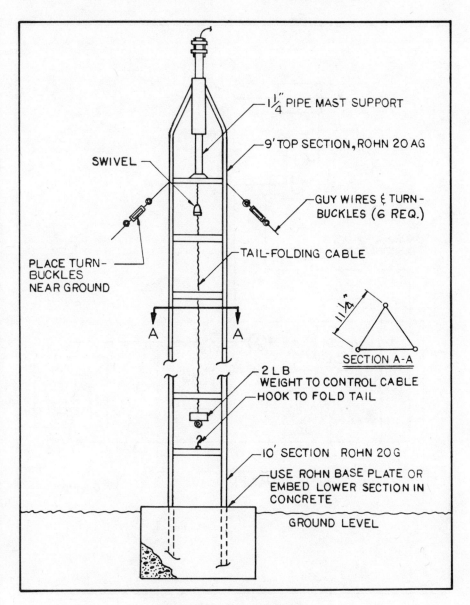

1¼" PIPE MAST SUPPORT

9' TOP SECTION, ROHN 20 AG

SWIVEL

GUY WIRES & TURN-BUCKLES (6 REQ.)

TAIL-FOLDING CABLE

PLACE TURN-BUCKLES NEAR GROUND

A   A

1¼"

SECTION A-A

2 LB WEIGHT TO CONTROL CABLE

HOOK TO FOLD TAIL

10' SECTION ROHN 20 G

USE ROHN BASE PLATE OR EMBED LOWER SECTION IN CONCRETE

GROUND LEVEL

FIGURE 11-15. TOWER

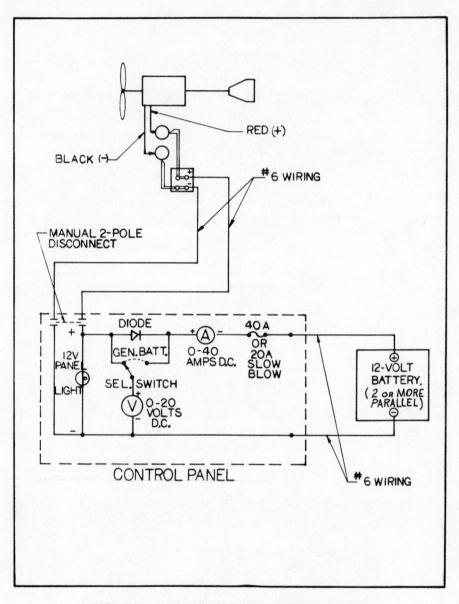

FIGURE 11-16. ELECTRICAL CONNECTIONS

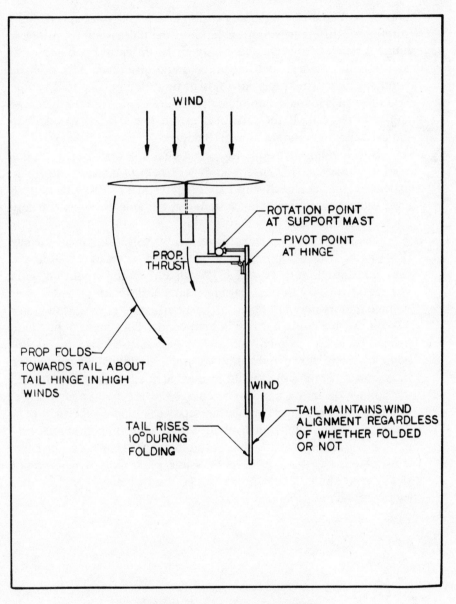

FIGURE 11-17. TAIL FOLDING

gravity pull created by the angled hinge at some wind speed. Substitution of different weight material in the tail assembly, different hinge distances from the supporting shaft, different pitch angles of the propeller blades, and tail hinge angle difference all cause the wind speed at which the tail folds to be different. The tail should fold in 35 MPH winds automatically. If it does not, reduce the hinge angle on the tail. If the tail folds at too low a wind speed, then increase the hinge angle above 10 degrees.

Bob Landing has gone to a great deal of work in creating this working system. This information will help you avoid the many expensive mistakes that any experimenter makes in achieving an end result such as this. This system can be built for approximately $350, plus batteries.

The motor recommended here is a 36-volt, permanent magnet (PM) field, with a continuous armature current rating of 20 amps. (See parts list Figure 11-4). A 12-volt PM motor would not gain voltage as quickly during starting, (same RPM rating) and higher voltage motors may not have sufficient armature current ratings for 12-volt use and could burn up in high winds. This does not preclude using 120 volt or higher voltage PM motors to charge a 120-volt battery system that would draw 6 amps armature current at one horsepower. PM motors should be rated at no more than 2 or 3 times the voltage of the batteries to be charged to get high enough armature current ratings to reach the horsepower rating on the motor. In this case, the recommended motor will not sustain continuous current above 20 amps without armature damage. However, intermittent operation to 40 amps is permissible since most winds are not steady in velocity. This unit was tested for 1½ years with no armature overloading problems.

12.

How to Make a Water Wheel

Before going into detail on the water wheel I tested, I will tell you how to figure the size of your water wheel. First you need to know what height you will be able to fall the water (head). In my case, I built a 2-foot dam in the creek. (Obviously, an overshot or pitchback water wheel was out of the question because it would take an 18-inch diameter wheel to get completely under the falling water. This would not hold sufficient water to power anything except a toy.) I consider a 3-foot diameter water wheel as the smallest use-able size. From a 2-foot waterfall, a 3-foot diameter water wheel will operate as a breastshot type. (See Chapter 6, Figure 6-1, for types and bucket shapes.)

What width should the water wheel be? This requires a calcula-tion and is dependent on the water flow in cubic feet per minute. Methods of calculating water flow (Q) are given in Chapter 6. As an example for this water wheel let's say you have available 45 cu. ft./min. of water coming down your water trough and falling onto the water wheel. You need to know the RPM of the wheel. This is a variable, but dependent on diameter and water flow. However, since more electrical load would be placed on a larger water flow, the major variable becomes diameter. (Figure 12-1 shows a table of approximate RPM's for diameters. These are not exact, but do fine for wheel width and motor RPM change ratio calculations.)

For our example the water wheel diameter is 3 feet, making the RPM 10 (from the table). With 45 cu. ft./min. of water flowing at 10 revolutions per minute, you get 4.5 cu. ft. of water per wheel

WATERWHEEL DIAMETER (FEET)	APPROXIMATE RPM FOR CALCULATIONS
3	10
4	9
5	8
6	7
7	6
8	5
9	4
10	3
12	2

FIGURE 12-1. TABLE OF RPM'S

revolution. I used 8 buckets on my wheel and this makes .5625 cu. ft. per bucket water flow. The cross sectional area of my bucket is 54 sq. inches or .375 sq. ft. Dividing .375 into .5625, I get 1.5 ft. for the width if I fill the bucket. Remembering that an efficient water wheel only fills the buckets about ⅓, the width should be 3 × 1.5 = 4.5 ft. I mistakenly did not allow for ⅓ full buckets on my unit and later had to rebuild it wider. The formula for the water wheel width is:

$$W = \frac{FLOW \times 3}{(RPM)(No.\ of\ buckets)(section\ of\ bucket)}$$

where W is in feet,

RPM is revolutions per minute from Figure 12-1,

No. of buckets—usually eight but could be more on large water wheels.

Section of bucket = ½ Depth × Width (if triangular) in sq. ft.

The water wheel tested is 3 feet in diameter, has eight buckets, is 18 inches wide, and turns at 10 RPM. It is made from redwood although a non-rusting metal may be used.

The unit uses a 3-belt pulley drive. For my motor, I used an 1800-RPM rated, 90-volt unit that I wished to drive at 300 RPM. I used a belt ratio of 30 to 1. (There is a discussion of motor speed and using DC motors as generators in Chapter 10. My interesting experiences in finding a suitable generation unit as described in Chapter 10 are worth reading if you are to avoid making the same mistakes.) In this chapter, I will tell you which motor I used as a generator and then recommend a better alternative. The following paragraphs give detailed plans on one specific water wheel that I built and used to power a small amount of lighting in two cabins in a remote area of the mountains in California where no commercial electricity exists.

DETAILED DESCRIPTION

The water wheel is a wooden 3-foot diameter wheel, 18 inches wide, with 8 buckets. (See Figure 12-2 for a photo of this water wheel.) It turns on a 1-inch solid steel shaft with ball bearings at each side. One side has a 12-inch diameter pulley attached to the shaft, driving a total of 6 pulleys and three belts to a ¼ HP motor at an RPM ratio of 30 to 1. The motor tested is a permanent magnet field DC motor rated at 120-volts and 3 amps armature current. A 36-volt motor would be better and is shown in the parts list.) It is charging a 12-volt battery system at 36 watts continuously. (See Figure 12-3 for the parts list.)

THE WHEEL

Building Instructions

First, the two circular sides are made by using five 1″ × 8″ redwood pieces in each of two directions, 90 degrees apart as shown in Figure 12-4. This method may be used for any size water wheel. After nailing the boards just enough to hold them temporarily together, draw a circle on one side using a nail in the center and a string from the nail to a pencil. After the marking is complete, put more nails in the boards being careful to avoid the circle you have to saw out. Saw the circle off with a sabre saw or rotary saw. I found the rotary saw easier on this 1½-inch thick cut even though I had to

FIGURE 12-2. WATER WHEEL

QTY   DESCRIPTION

WHEEL

QTY	DESCRIPTION
20	1″ × 8″ × (length equals diameter) redwood
24	1″ × 8″ × (width of wheel) (1″ × 6″ may be substituted) redwood
2 lbs	10d nails
24 ft.	Angle material, ¾″ × ¾″, (aluminum or sheetmetal)
8 doz.	Machine screws, #8 × 1¼″
8 doz.	Machine screws, #8 × 2″
32 doz.	Flat washers, #8
16 doz.	Lockwashers, #8
16 doz.	Nuts, #8
8	L brackets
1	Shaft, 1″ dia. × 3′ long, cold roll steel
2	1″ pipe flanges
1	1″ pipe nipple, 3″ long
8	Wood screws, #10 × 1½″ long
4	Set screws

WATERWHEEL BEARINGS AND SIDE BOXES

QTY	DESCRIPTION
1	4′ × 8′ sheet ½″ thick A-C plywood
24 ft.	2″ × 6″ fir
24 ft.	2″ × 4″ fir
1 set	Hinges, hasp, and lock (for pulley box door)

PULLEYS, BELTS, BEARINGS

QTY	DESCRIPTION
2	1″ dia. ball bearings, w/4 hole flange and grease fitting
2	½″ dia shaft × 7″ long, cold roll steel
4	½″ dia. ball bearings, w/4 hole flange and grease fitting
8	⅜″ dia. wood screws w/hex head, 2″ long (for 1″ bearings)
8	$5/_{16}$″ × 3½″ long hex bolts w/nuts (for ½″ bearings)
1	Pulley, 12″ dia. for 1″ shaft
1	Pulley, 3″ dia. for ½″ shaft
1	Pulley, 11″ dia. for ½″ shaft
1	Pulley, 3″ dia. for ½″ shaft
1	Pulley, 9″ dia. for ½″ shaft
1	Pulley, 3″ dia. for generator shaft
3	½″ wide V belts, Type A, length as measured on pulleys installed

GENERATOR & ELECTRICAL

QTY	DESCRIPTION
1	Motor, DC PM field, 36 volts, 1800 RPM, ¼ HP
1	Diode, voltage rating twice battery, current rating twice armature current rating of motor.

FIGURE 12-3. PARTS LIST FOR WATER WHEEL

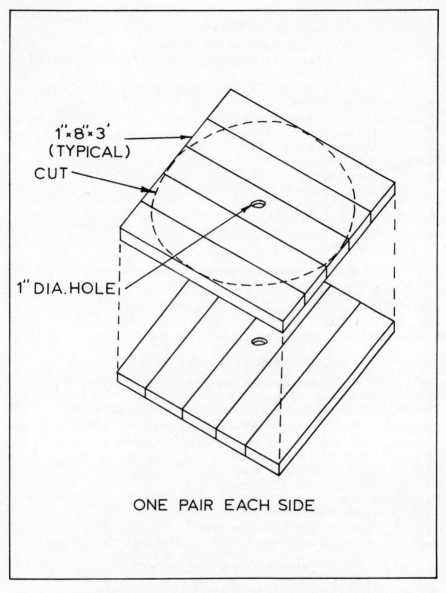

1"×8"×3'
(TYPICAL)

CUT

1" DIA. HOLE

ONE PAIR EACH SIDE

FIGURE 12-4. WHEEL SIDES CONSTRUCTION

make many slices off the edge. Once started, the rotary saw can be turned along about a 2-inch circumference before going straight off the edge. After sawing, mark the locations for the buckets on the inside of each wheel. Keep in mind that the bucket marks are reversed on one side so that they match when placed face to face. (See Figure 12-5 for marking dimensions for this wheel.)

Next drill a 1-inch diameter hole through the center of each side (at the previous nail location you used to make the circle). A high speed wood bit for electric drills is great for this task.

Next put the buckets onto the sides, using angle material and machine screws as shown in Figure 12-6. The angle for a small water wheel like this one can be made from wall board cornering material. This metal is very thin and should not be used on larger wheels. Use L-brackets at the center of each bucket to strengthen it there.

The shaft is 1-inch diameter, cold roll steel 3 ft. long (18 inches longer than the inside wheel width). Insert this through the center of the water wheel. You should have it slightly tapered at the ends of a grinding wheel to insure a sliding fit through the bearings. The shaft should stick out on the pulley side 9 inches, and 6 inches on the other side. Fasten the shaft to the wheel using a 1-inch pipe flange, and a 3-inch long, 1-inch pipe nipple. The nipple is cut in half using a piece on each side to screw into the flange. The flange is attached to the wheel using four #10 × 1½-inch wood screws on each flange. Before screwing in the nipple halves, drill two holes in each one and tap for machine screws to be used as set screws. (See Figure 12-6.) Then screw the nipples into their final resting position. Use a punch through the tapped holes to mark the 1-inch shaft. Remove the nipples. Drill ⅛" deep holes at the punch marks on the shaft. Make the holes bigger in diameter than the set screws. Now replace the nipples lining them up with the drilled holes. Put in the set screws and paint or silicone seal them in place.

WATER WHEEL BEARINGS AND SIDE BOXES

The water wheel bearings, box, and pulleys are arranged approximately as shown in Figure 12-7. This figure is meant to show general clearances for bearings and pulleys. In the figure, one pulley

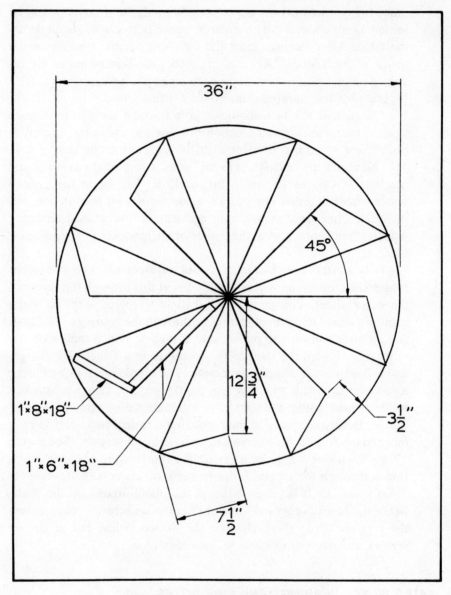

FIGURE 12-5. BUCKET LAYOUT

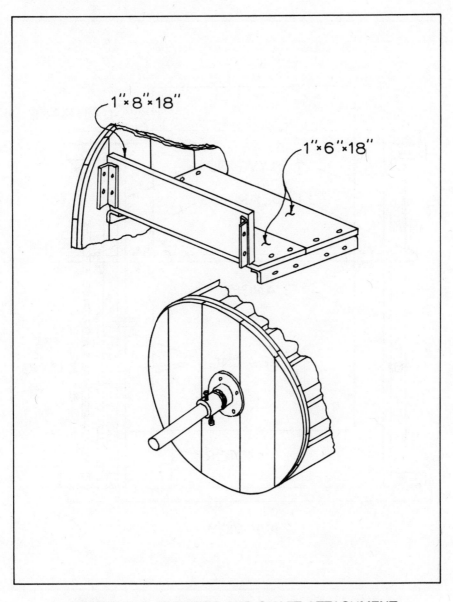

FIGURE 12-6. BUCKETS AND SHAFT ATTACHMENT

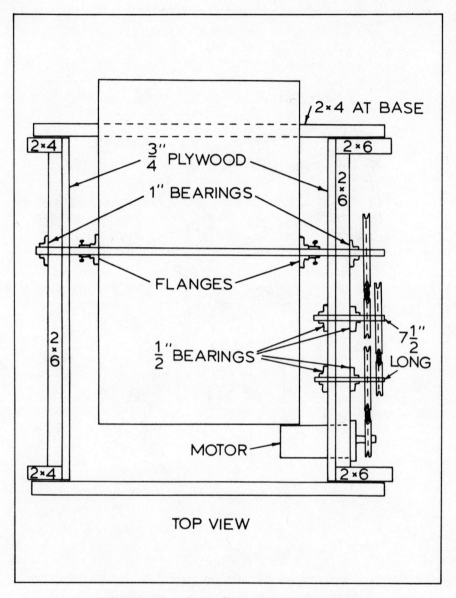

FIGURE 12-7. TOP VIEW ARRANGEMENT

shaft has been displaced away from the main shaft for clear illustration of the principle of operation. (See Figure 12-8 for pulleys and pulley box layout.) The water wheel side of the box is covered with ¾ inch thick A-C plywood, 28″ × 36″. A thickness of ½-inch may be substituted. The front of the pulley box gets a hinged door. The two ½-inch pulley shafts in the middle are made adjustable by slotting the holes for the shaft and bearing-to-bearing bolts in the direction of the arrows in Figure 12-8. (Also see Figure 12-9.)

The non-pulley side is made from 2 × 4's except for one 2 × 6 under the 1 inch bearing. Connect the two boxes with 2 × 4's or 2 × 6's to clear the water wheel and motor access.

The motor is installed with its flange on the inside of the box because otherwise the shaft won't reach through the 2 × 6. If the door is locked this will help reduce theft of the motor. Make the hole through the 2 × 6 just big enough to accept the motor housing using a notch for wiring or other protrusions.

The permanent magnet field DC motor that I used is good for 3 amps output. (If you want more, use a higher current, lower RPM, PM DC motor from the parts list in Figure 12-3 and the company source listed in the Figure 11-4 parts list for the motor.) (For wiring see Figure 12-10.) Usually, a red and black lead come out of the motor. Clockwise rotation usually causes positive voltage on the red lead. Check this with a voltmeter to be sure before connecting it. This design rotates clockwise and provides positive on the red lead. Run the two wires to the battery. Connect the black lead to the battery negative bus. Connect the red lead through a diode with a good heat sink to the positive battery bus. The diode polarity is shown in Figure 12-10 and should be rated for twice the battery voltage and twice the armature current. This diode will allow your motor to operate as a generator and prevent reverse current which would run the motor from battery power.

This concludes the water wheel information. If it is not what you need, design your own from the notes in the beginning of the chapter because water wheels are quite easy to design.

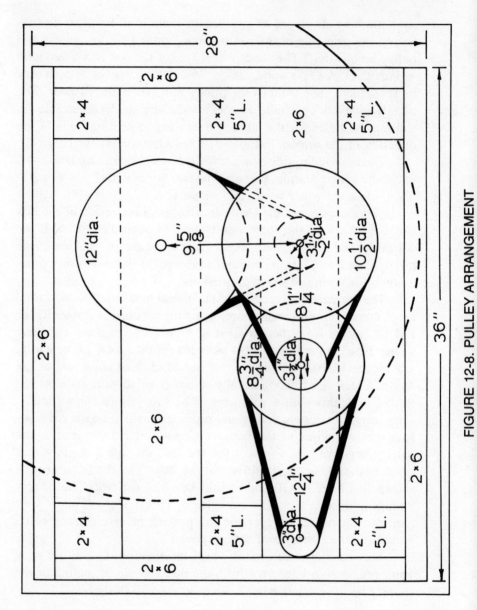

FIGURE 12-8. PULLEY ARRANGEMENT

FIGURE 12-9. PHOTO OF PULLEYS

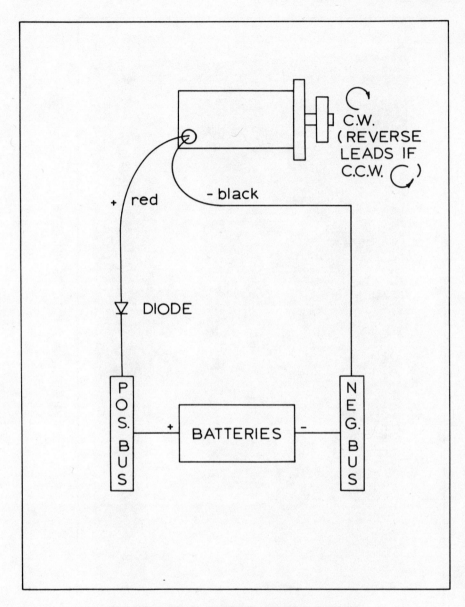

FIGURE 12-10. WATER WHEEL WIRING

13.

How to Build a Bicycle Exerciser

DESIGN GUIDES

I mentioned in an earlier chapter that the human body in bicycling is capable of about ⅓ horsepower. This is dependent on body size and personal physical condition. To elaborate on this point, let me give you some examples. Some professional bicyclists have produced 1 horsepower for short periods. The winner in a 12-hour bicycle race produces an average power output over the 12 hours of ⅓ horsepower. In a 24-hour race, the winner can average ¼ horsepower. An untrained, middle-aged bicyclist can produce $1/10$ horsepower for one or two hours without undue fatigue. A Dartmouth College student showed that the average utility cyclist produces $1/20$ horsepower at 8 miles per hour.

The muscle group that does all this is the quadriceps thigh muscle, the largest and most powerful in the body. The trick is to put it to work in an exerciser that is cheap, simple to build, and adjustable in load. Since several members of your family may wish to use the exerciser, an adjustable load is essential and is also extremely simple to provide. Most bicycles today have gear-shift levers to attain 3-speed through 10-speed gearing. Why? To adjust for load variations such as uphill, downhill, and speed on flat terrain. Without modification, a bicycle gear shift can be used to vary the speed of a generator, thereby varying the load and power output for different people. Furthermore, there is no need to modify the bicycle used in any manner. In fact, only two things are necessary to make an exerciser generator from a bicycle: (1) generator drive from the rear bicycle tire; and (2) a means of holding the bicycle up during the exercise. These can be made as a single unit with the bicycle removeable for normal outdoor use. In this case, only two electrical wires (to the ammeter) must be removed to remove the bicycle from the exerciser.

DESIGN NOTES

Some calculations are necessary to arrive at the design. Assuming that anyone can make a bicycle go 10 miles per hour (MPH) given a gear-shift choice, I used this as a near maximum RPM for the rear wheel of the bicycle. Even though some people can make it go twice as fast, they can't do it uphill. Since uphill more closely simulates the load we are thinking of, the 10 MPH figure makes a nice design goal for top generator RPM. Converting 10 MPH to feet per minute (multiply by 88), we get 880 feet per minute. This is the design top speed of the bicycle rear tire surface. Any wheel pushed against the rear tire of the bicycle will have the same surface speed. The design will be for a generator running at 1200 RPM and the wheel that presses against the bicycle tire will be attached directly to the generator shaft.

The size of the wheel needed is calculated as follows. The circumference of the wheel is

$$\pi \text{ D } (3.1416 \times \text{diameter}).$$

One circumference equals one revolution. Surface speed in feet per minute divided by revolutions per minute yields feet per revolution ($880 \div 1200 = 0.73$). If there is 0.73 feet in a revolution, then the circumference equals 0.73 feet on the wheel we are looking for. The diameter of that wheel is

$$0.73 \div \pi = 0.73 \div 3.1416 = .23 \text{ feet} = 2.8 \text{ inches diameter.}$$

This is only an approximation because of our 10 MPH starting assumption. I used a 2½-inch diameter wheel because it was handy to make on a lathe. No belts or gearing are required except that gearing already built into the bicycle.

The horsepower rating and type of generator are the next decisions. For various reasons, (outlined in Chapter 10 in detail) alternators have been ruled out for this application. The primary reasons being that the first $1/20$ horsepower is used in the field current, and the total size of most auto alternators approaches 1 horsepower which is too much load for an exerciser. Other disadvantages are RPM (for most alternators 4000 RPM) requiring between 6 and 10-to-1 belt ratios and additional pulleys. The horsepower rating for the needed generator should be between $1/10$ and $1/20$ horsepower at 600 RPM. The voltage rating should be similar to the battery you are charging. I recommend using a permanent magnet field DC motor as a generator. A diode in the positive output line will prevent any motoring action. These units make excellent generators for this purpose.

Wanting to keep my costs down, I purchased two $1/20$ horsepower PM DC motors, used, for $6 each. I attached these, one to each end of the pulley shaft; the pulley rides against the rear tire of the bicycle. One motor turns clockwise (CW) and the other turns counterclockwise (CCW). The one that turns CCW must have its output leads reversed but otherwise the motor doesn't care. These $1/20$-HP PM DC motors are often used as electric window motors in automobiles and may be picked up used, from salvage yards or electrical used parts stores. The combination of the two motors puts out 12 volts at 6 amps = 72 watts (about $1/10$ horsepower). (For your reference: 1 horsepower equals 746 watts and volts times amps equals watts on a DC system.)

DETAILED DESCRIPTION

This unit is so simple to build that you may find it a one day task. The bicycle of your choice is fitted into slots on the exerciser framework. The framework raises the bicycle about six inches off the floor. Therefore, a girl's bicycle is recommended for easy entry to the seat. The exerciser framework holds the front and rear wheels upright via axle slots. Two springs hold the hinged motor-pulley mechanism against the rear bicycle tire. The two generators are connected in series via a diode and ammeter to the battery to be charged. (See Figure 13-1 for a photo of the system. See Figure 13-2 for the parts list.)

FIGURE 13-1. BICYCLE EXERCISER

QTY	DESCRIPTION
16 ft.	2″ × 4″ fir
8 ft.	2″ × 6″ fir
1 lb.	10d nails
4	Metal plates (to rest axles on)
2	Hinges
2	Springs (adjust for minimum tension)
1	3½″ dia. aluminum pulley with groove for tire lathed down to 2½″ dia. (1¼″ wide)
2	$1/20$ HP motors (or one $1/10$ HP motor) DC, with permanent magnet field, rated 12 volts, 600 to 1800 RPM.
1	Ammeter, 10 amps DC with brackets
1	Diode, 12 amp, 24 volt, with heat sink
1	Girl's bicycle, 3-speed or up to 10-speed

FIGURE 13-2. EXERCISER PARTS LIST

BUILDING INSTRUCTIONS

See the photographed details in Figures 13-3, 13-4, 13-5 and 13-6 for complete detailed construction. Step by step instructions will not be given. The following notes should be adhered to for problem free construction.

1. Your support frame must be very flat and stable where it rests on the floor.
2. The bicycle you use will vary from others in front axle to rear axle distance, axle length, and other characteristics important to this project. So use the exact bicycle you will be using when fitting the bicycle support during construction.
3. Different generators will require some design on your part to mount them into the 2 × 6 frame with free-spinning pulley attached.
4. The springs used to hold the motor pulley against the bicycle tire should be just strong enough to prevent the pulley from slipping on the tire. Any excess pressure on the pulley will result in noticeable friction losses in motor bearings and reduced bearing

FIGURE 13-3. AXLE SUPPORT DETAIL

FIGURE 13-4. HINGED SPRING-LOADED MOTOR PULLEY GROUP

FIGURE 13-5. MOTOR-PULLEY ATTACHMENT AND SPACING

FIGURE 13-6. AMMETER ATTACHMENT

life. The wood screw type hooks for the springs near the motors have had the pointed end blunted by filing so that they act as a set screw against the motor housing preventing each motor from turning in its mounting.

5. Stability problems, should they occur on the bicycle, can be eliminated by an A-frame structure to the front steering column. I had no stability problems.

6. Excess sizing of the generator will cause loss of interest in exercising resulting in no battery charging. Construct small ($^1/_{10}$ horsepower)!

7. See Figure 12-10 for correct use of diodes with PM DC motors charging batteries.

8. Remember the bicycle axle's weight must be on your framework. To prevent wood indentation with use, use a metal side plate for axle contact, making a groove in the metal for the axle to rest in.

It cost me less than $50 to make this exerciser. In fact, since I had most of it laying in my junk pile, I only purchased $6 worth of springs and hinges. You may not be so fortunate, but you can stay under $50. This price assumes that you already have a girl's bicycle.

14.

*Mirror-Target Systems**

In Chapter 5, a discussion of the sun and photovoltaic methods for extracting electricity were discussed. In this chapter, I will make specific application notes on mirror-target systems. These systems are heat-to-motion converters where the motion is used to turn generators. A rough design for such a system is presented in this chapter. Most of the industrial work done in this field so far has been for utility-owned power plants, several of which are in the wings of the Federal Energy Research and Development Administration (ERDA) for feasibility studies and pilot plants. As the 10 MW (megawatt) pilot plants prove themselves, larger solar-electric power plants will be built, probably in the 100 MW range. It only takes time and money. Most of these experimental plants use multiple mirrors, reflecting energy on a tower receiver. The tower receiver operates as a boiler (usually, or as a superheater) to drive a steam turbine-generator (other fluids or gases may be used in lieu of steam).

SYSTEM DESCRIPTION

Size is important in considering any system, since it affects almost all of the design. The discussions here will be based on that amount of power required for a single dwelling. Dwelling size varies

---

*This is the one system in this book that has never been built by me (because of cost) nor by anyone else either but because it is the largest experimental source of sun power, basic design considerations are presented for those interested in learning or experimenting. No detailed design or building instructions are provided nor intended.

from apartments requiring as little as 10 kilowatt hours (KWH) per day to large homes requiring as much as 70 KWH per day. For design purposes, I picked 25 KWH per day.

The 25 KWH per day is based on a heavily insulated, 1500-square-foot house, all electric, with solar heating and cooling, and exact calculations on electrical loads. I will build the house for myself shortly, and it should closely resemble needed loads for houses of the future.

The basic system considered here is a mirror field, reflecting energy onto a boiler. The boiler produces steam to run a steam engine. The steam engine is attached to a permanent magnet field, DC generator by belt or shaft to create 125 VDC power output. The DC power charges a battery bank of sufficient ampere-hour storage to handle the number of cloudy days expected at one time. The battery power is inverted to 3-wire, 220/110 volts AC for normal electrical loads as used today. Automatic switching will allow use of commercial power as back up should batteries not create sufficient storage (as happens occasionally) or for private power component failure. (See Figure 14-1 for a system diagram.) Methods of accomplishing each task in the system are described and illustrated in this chapter. This is not the only way to design such a system, but it is one practical way. For the sake of simplicity, subsystems that improve efficiency have been left out of Figure 14-1. However, the methods and principles of improving efficiency by certain subsystems will be discussed.

MIRROR FIELD SIZING

The mirror field will be made up of hundreds of small, flat mirrors, each reflecting the sun's energy (insolation) onto the same target boiler. The first question that must be answered is what is the total surface area of mirrors required?

A solar-heated and heavily insulated single dwelling requires about 25 KWH per day of electricity. To find your daily needs, divide your average monthly KWH's used by 30.4 (average days in a month) or your annual KWH's used by 365.

One square meter of mirrors (about 39" × 39") will receive about 1000 watts peak power at noon solar time if normal to the sun's rays. The power lasts for about 10 hours (less in winter, more

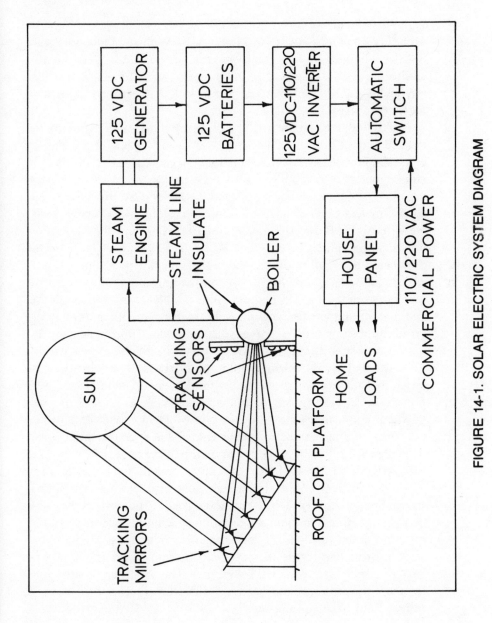

FIGURE 14-1. SOLAR ELECTRIC SYSTEM DIAGRAM

in summer) and averages about 800 watts per square meter over the 10 hours on clear days. (See Figure 5-3.) With reference to Figure 5-4 in Chapter 5, the mean annual sunshine for Sacramento, California is 3400 hours. Multiplying by 800 watts per square meter, yields 2,720 KWH per year (or 7.45 KWH per day) per square meter (average).

To find the square meters of mirrors required, start at the desired output and work backwards. I want 25 KWH per day of electrical power. At 7.45 KWH per day per square meter this is 3.36 square meters of mirrors (at 100% efficiency). Since we are expecting only 25% efficiency, we need (3.36 divided by 25%) 13.4 square meters (M^2) of mirrors. Since the efficiency is a best guess, we will round off the size upward to 15 M^2. For ease of building, we will pick the dimension of 3 meters × 5 meters (9'9" × 16'3"). This dimension does not allow for individual mirror separation and possible shadowing of one mirror on another. Assuming that no mirror will move more than 45 degrees from parallel to its support frame and that the support frame will always be between 30° and 60° to the ground plane, it can be proven that individual mirror spacing required to insure minimum shadowing is 58% of the mirror width.

The individual mirror width should be selected by the size target that the boiler can withstand in a given available target area. For example, our 15 M^2 of mirrors will receive 15 KW of peak power at noon solar time if normal to the sun and about 85% of this will reach the target due to mirror angles and specular reflectance. This places 12.75 KW peak power on the target area (1 KW = 3413 BTU/HR) or 43,516 BTU/HR. If a target area on the boiler of 3" by 3" (9 sq. inches) could accept that heat level within operational range, then mirrors of 3" × 3" could be used. (The mirrors may also be rectangular and in that case required spacing of mirrors would be different in the vertical and horizontal directions.) Assuming that 3" × 3" mirrors and target size are OK, the mirror spacing would be 58% of 3 inches (1¾") in each direction. The backframe dimensions for the entire mirror field becomes 58% larger. The 9'9" × 16'3" mirror field is spread out over a 15'3½" × 25'7" framework.

MIRROR MOUNTING

Each mirror must be mounted so that it can change its position both in azimuth and elevation. Since the sun can elevate more than 90° within 23½° of latitude from the equator, and can change azimuth by 227° in mid-summer, one might think that mirror tracking angles must make these changes also. However, flat mirrors reflecting energy onto a stationary target need only move half the angles the sun does because these mirrors must always point to a position half way between the target and the sun.

In Figure 14-2, the incident angle (theta) is equal to the reflectance angle (theta) on a flat mirror surface. Therefore, to enable the reflected energy to strike the target, a line perpendicular to the mirror surface must always bisect the angle between the sun and the target. This means that if the sun is 90° from your target, the mirror must only move 45° off target to reflect that energy onto the target. (The actual sun angles vary during the year. See Figure 14-3 for an illustration of the sun's relative movements during the year.)

In Sacramento, California the latitude is 38½° N. In this case, the sun will reach 28 degrees above the horizon in mid-winter and 75 degrees above the horizon in mid-summer. Assuming your target is just below the winter sun at 7 degrees above the horizon (this will be explained later), your mirrors (at noon) would point to $\left(\dfrac{28-7}{2}\right) + 7 = 17\frac{1}{2}$ degrees above the horizon in mid-winter and $\left(\dfrac{72-7}{2}\right) + 7 = 41$ degrees above the horizon in mid-summer. The annual average elevation angle of the mirrors would be $\dfrac{17\frac{1}{2} + 41}{2} = 29\frac{1}{4}$ degrees. Therefore, a mirror backframe angle of 60¾ degrees (90 − 29¼) to the ground is indicated for the best mirror elevation performance (pointed south). However, this makes azimuth corrections difficult. Therefore, a compromise will be made to a backframe angle of 30 degrees above the ground. Azimuth corrections will be made along the backframe angle. This means that each azimuth correction causes an elevation error and each elevation correction (when not pointing south) will cause an azimuth error. This is acceptable in a closed loop control system since the errors will be sensed and corrected.

In general, a single mirror mount must rotate from south ± 57° and elevate ± 30° from 30° above the horizon. This can be accomplished mechanically as shown in Figure 14-4. To simplify controls, we will limit azimuth to ± 45 degrees. This will prevent

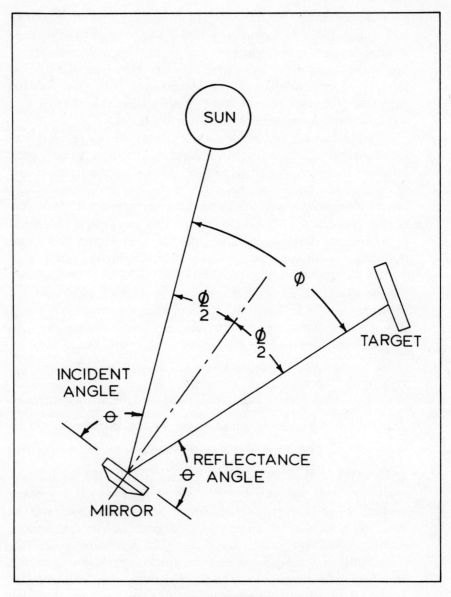

FIGURE 14-2. MIRROR ANGLE

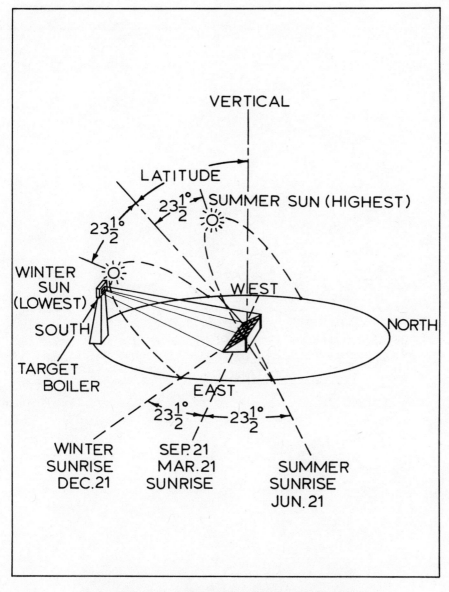

FIGURE 14-3. SUN'S RELATIVE MOVEMENTS
(NORTHERN HEMISPHERE)

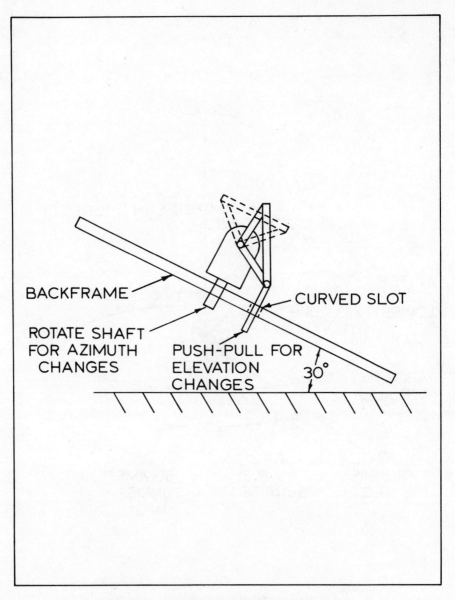

FIGURE 14-4. SINGLE MIRROR MOUNT

near horizon summer tracking but the mirrors are shadowed by each other and available insolation is negligible at this time of day anyway.

How the mirrors mechanically relate to each other is a function of control.

MIRROR TRACKING AND CONTROL

There seems to be a common misconception that with multiple flat mirrors in a field that all follow different tracks requires exotic computer control. The plain truth is that, although each mirror points in a different direction at the beginning and end of tracking corrections, the angle of change and rate of angular change are identical for all mirrors. This means that once the mirrors are aligned to reflect energy onto the same target, they may be locked together and given the same tracking information for all tracking. (See Figures 14-5 and 14-6 for a simple proof.)

Both azimuth and elevation changes for every mirror are *always* equal to ½ the angle of change of the sun over *any* period of time. Corollaries of the proof show that it is valid for any mirror location so long as the mirror remains at a fixed distance and fixed angle to the target receiver. So the mirror field can be slanted, bowed, or whatever as long as each individual mirror rotates on its own axis.

The locking of all mirrors together for tracking information (after alignment) may be accomplished mechanically or electrically. With a system of this size, it can be accomplished mechanically. Larger systems will be stuck with electrical interlocking of some kind. With mechanical interlocking, only one reversible azimuth tracking motor and one reversible elevation tracking motor are required.

Since the tracking information will be the same for all mirrors, closed loop sensing need not check individual mirrors. Sensors will be placed around the target boiler to sense off-target heating for the entire mirror group. Since these sensors must withstand high temperatures, type J thermocouples (good to 1400°F) will probably suffice. To make the voltage signals easy to work with, dual quadrants of thermocouples in series (thermopile) could be used. These thermocouples would be mounted in four groups, each covering half

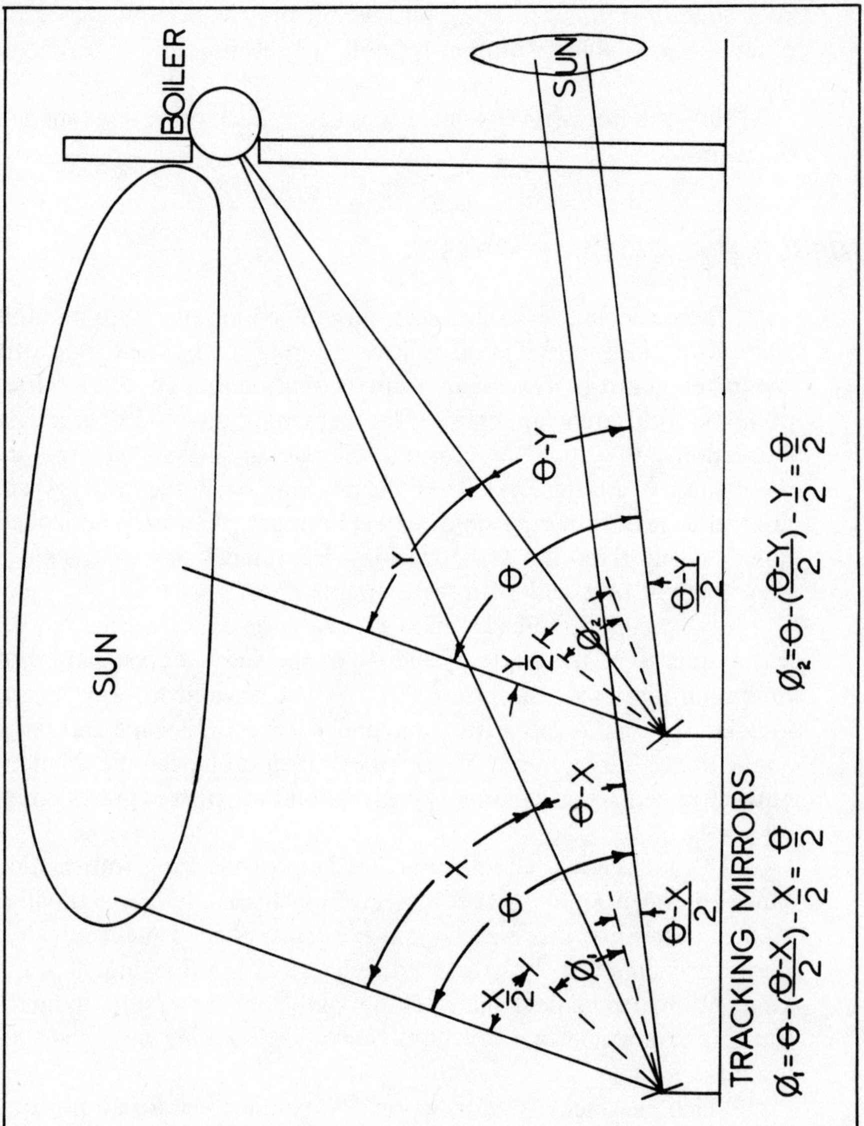

FIGURE 14-5. FUNCTIONS OF ANGLE CHANGE (ELEVATION)

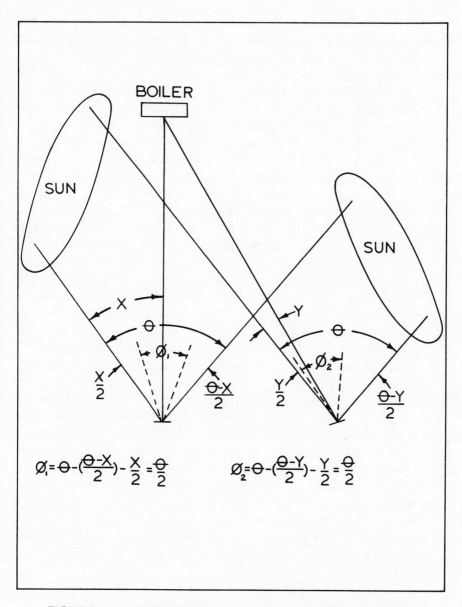

FIGURE 14-6. FUNCTIONS OF ANGLE CHANGE (AZIMUTH)

of the off-target area. Down elevation sensors will occupy all the off-target area above the target center, and up elevation sensors will occupy all off-target area below the target center. The two azimuth sensor groups will be to the left and right of the target center for their respective signals and will overlap the sensors used for elevation signals. (See Figure 14-7.) The off-target board containing the thermocouples will be made of nonflammable material and be of such size that vertical errors of 14 degrees and horizontal errors of 23½ degrees will be picked up for correction. This will allow limiting azimuth tracking of the mirrors to ± 45 degrees.

The closed loop control scheme is shown in Figure 14-8. It is self-explanatory. Not shown, is a light sensor which provides contact closure to drive the azimuth and elevation motors to the "dawn" starting position after dark. Limit switches stop the motors when they reach the dawn starting point. A single starting point is provided for year-around use. It assumes the sun will always be due east at the horizon. But the sun will deviate ± 23½° in azimuth from summer to winter as shown in Figure 14-3. Therefore, the sensors must cover azimuth changes of ± 23½° as shown in Figure 14-7. In mid-summer, the mirrors are already at 45° deflection (90° sun) when the sun rises at 113½° azimuth. Vertical tracking only may take place until the sun passes due east, then azimuth tracking begins as sensor signals take the mirrors off the 45° stop location.

In summary, we have a tracking system that provides a single set of tracking signals for the whole mirror field and single starting point for tracking. It is believed that this system will track in light clouds and fail to track in heavy clouds. If left unattended, the system will return to start each morning and could lose afternoon sun after a cloudy morning. A manual control should be included to bring mirrors onto target in midday or a timer could be employed to sweep once an hour if no signals are received. I prefer manual backup for starting tracking in midday so that a simple system can be retained.

TARGET BOILER

The target boiler must be located so that it and its sensor field are just below the mid-winter noon sun, as illustrated in Figure 14-3. That angle is measured from the bottom of the lowest mirror

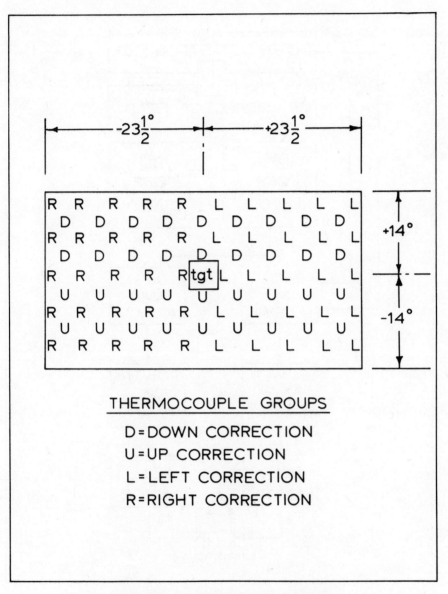

FIGURE 14-7. SENSOR LOCATIONS

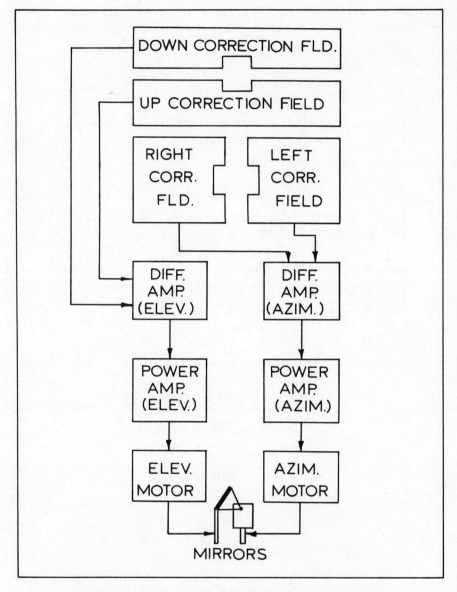

FIGURE 14-8. CLOSED LOOP CONTROL SCHEME

across the top of the sensor field and must be less than 90 degrees minus 23½ degrees minus your latitude. (See Figure 14-9 for Sacramento calculations.) That angle is 28 degrees in Sacramento and I divided it in half for the elevation sensor fields. With an available angle of 30 degrees or more, use ± 15 degrees for the elevation sensors field. The target should be at least as far away from the mirrors as the width of the mirror field. This will guarantee an angle of energy convergence on the target of less than 60 degrees. This is necessary to keep the target opening as small as possible thereby reducing heat losses from the target.

The boiler may be a flash boiler or the water-filled kind. It should be designed to handle the BTU/hour input (divided by 976 equals pounds of steam per hour). Boilers are rated in pounds of steam per hour. With 43,516 BTU/hr available, this is 44.5 pounds of steam per hour maximum. It will be difficult to locate a boiler this small and it will have to be built. Boiler design is a science in itself. Consult a boiler maker to obtain one. Insulate the boiler for minimum heat loss.

STEAM ENGINE

Small steam turbines are not available on the open market and are probably too high in RPM, maintenance, and cost for this purpose. Therefore, a steam engine is appropriate. You need a steam engine capable of driving a generator at 25 KWH per day. Assume a 5-hour day in winter, that is 5 KW's required continuously from the generator. Dividing by 746, we get 6.7 horsepower. Your steam engine must create 6.7 horsepower divided by the efficiency of the generator (about 85%) equals 7.9 HP. The piston size and operating pressure are dependent on RPM and power. The lower the pressure, the lower the loss in the steam engine. Aim for less than 600 RPM and 15 PSI operation. This will give you steam at 240°F or above. Lower temperatures have lower heat losses. If you can't find a suitable steam engine on the market, several build-it-yourself steam engine kits are available. Try the following sources for steam engines. Eight HP is about right.

AUTOMOTIVE STEAM SYSTEMS (steam engines)
8591 Pyle Way
Midway City, Calif. 92655

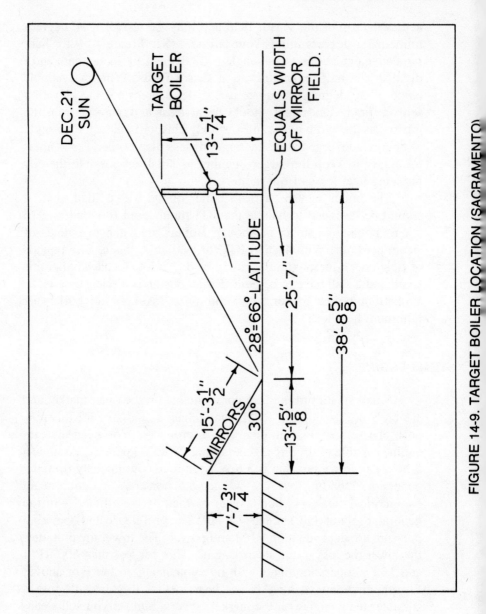

FIGURE 14-9. TARGET BOILER LOCATION (SACRAMENTO)

SEMPLE ENGINE COM. INC. (Steam engines, boilers, accessories)
Box 8354
St. Louis, Mo. 63124

RELIABLE INDUSTRIES, INC. (Steam engines, 4 HP to 200 HP)
(boilers, condensers, feed pumps and accessories)
34403 Joel St.
New Baltimore, Mich. 48047

Most of these vendors charge a dollar or so for their catalogue, but the price is worth spending as many provide design parameters.

GENERATOR

The generator should be a PM field, DC motor rated at 600 RPM and battery voltage. 125 VDC is recommended to keep current levels down. Although 6.7 horsepower is needed to obtain 5 KW output, you will find it difficult to find that exact horsepower. Get as close as you can on the high side (7 HP, etc.). A 5 KW output is 40 amps at 125 VDC. You will need a diode in the positive output line to prevent the DC generator from running as a motor. The diode should be rated 50 amps minimum, at 250 volts minimum.

BATTERY STORAGE

Battery storage should be 125 VDC, which is ten 12-volt batteries in series. The number of ampere-hours should be equal to your daily needs times the expected winter continuous cloudy days. If your needs are 25 KWH per day, at 125 VDC that becomes 200 ampere-hours (AH) per day (KWH divided by volts). For the Sacramento area, I would use 3 days for a total battery capability of 600 AH. (See Chapter 7 for a discussion of battery selection.) You should be able to purchase this 75 KW hours of storage batteries for about $2000.

INVERTERS

Run your entire household off of one or more inverters. For a well-insulated house with economical solar heating and cooling, three 2 KW inverters will do the job. But if you have poor insulation, a refrigerator type air conditioner, and an electric range and oven, all of which could contribute to the electrical load simultaneously, you will need 10 to 12 KW's of inverters for peak loads. I have found that 2 KW inverters give you the most power for your money. Therefore, multiple 2 KW units are the least expensive.

The specifications on a 2 KW inverter should read: (Ref: Nova Model 2K 60-150)

INPUT: 105-150 VDC @ 17.8 amps

OUTPUT: 117 VAC, 60 HZ, single phase

POWER: 2000 VA (same as 2 kilowatts at power factor 1.0)

APPROXIMATE COST: $2,500 each

EFFICIENCY: 75%

WEIGHT & DIMENSIONS: 200 lbs. 10½" H × 17¼" W ×
                                       20⅛" D

Try purchasing it from: Nova Electric Mfg. Co., 263 Hillside Ave., Nutley, N.J. 07110.

You will want a static switch to automatically switch commercial power to your house electrical panel when battery voltage falls below that necessary to operate the inverters (usually at 105 volts). Try to obtain this from your inverter supplier or a vendor that sells Uninterruptable Power Supplies (UPS).

SYSTEM CONSIDERATIONS

Some discussion of power plant efficiencies is appropriate. Expect 5 to 15% loss of energy at the mirrors due to mirror quality. Expect another 15% loss due to mirror angles not normal to the sun. Your insulated boiler and steam pipes will probably lose another 25%. You should be able to recover a large part of your used steam heat (more on this). Your steam engine will lose another 50%. Finally the DC generator will lose 15%. This gives you 85% left at the mirrors; 72¼% left at the boiler; 50% left at the steam engine; 27% left at the generator, and 23% going into the batteries.

This is normal, and assumes that you have recovered most of the heat in the used steam. Although exhaust steam in a large power plant would be condensed and reused with make-up water as appropriate, such subsystems are not economical on this size system. Furthermore, with steam engines, some engine lubricant oil will leak to the used steam and removing that oil would be mandatory for reuse of the water. Since pure water is a must for boiler and steam engine preventative maintenance, it is recommended that the water input be tap water with a high degree of mineral filtering (2 or 3 filters if necessary). This will prevent boiler and engine deposits from building up which will reduce your efficiency. This input water should be preheated by running the exhaust steam through a heat exchanger to pass as much of the heat as possible back into the system.

You may find more or less efficiencies in certain areas than the ones I mentioned. Experimentation is required to find exact efficiencies and this has not yet been accomplished. Your direct attack on any low efficiency section will result in an overall system near 25% efficiency. Power plants of this size have never before been built. It is estimated that this 5 KW power plant could be built for $10,000 including 75 KW hours of batteries and 6 KW of inverters. That's material cost only. Add labor and automatic switching for a full price.

There are some who believe that small power plants such as this at the user's location would greatly reduce the peak load demands on commercial electric systems. This design is untested, but is presented to gain your interest in experimenting. Before proceeding to put one of these systems on your roof or in your yard, check your building restrictions and local codes for boiler operation. Boiler operations usually need licensing by local authorities.

15.

System Connections, Maintenance, and Evaluation

INSTALLING BATTERIES

The physical installation of batteries is discussed in this chapter including connectors, battery boxes, future additions, and environmental conditions. (See Chapter 7 for ventilation and code requirements.)

The location of the storage batteries should be selected so that the batteries remain dry, are free from freezing or extremely hot temperatures, and are well-ventilated so that gases do not build up to explosive levels. Your site selection should also take into account accessibility for routine maintenance, replacement, battery filling and later battery additions.

The equipment shown in Figure 15-1 is usually located at the batteries. These include the batteries, the buses, terminal blocks, diodes from incoming power, and shunts for ammeter instrumentation. The batteries should be connected as shown in Chapter 7.

Figure 15-1 shows a 12-volt battery system with five batteries in parallel. (For series connections see Chapter 7. Do not use parallel systems for 125 VDC or large storage needs.) Each battery is connected to the negative and positive bus with standard auto-type battery cable (illustrated at the upper left battery in Figure 15-1). On the bus end, cable connectors with ¼-inch holes, minimum, are provided and are bolted to the bus with ¼-inch or $^5/_{16}$-inch diameter × 1-inch long bolts. The buses are made from copper or aluminum bars 1-inch wide × ¼-inch thick, by length as necessary to span batteries plus future locations of batteries. Every 1-inch to 1½-inch

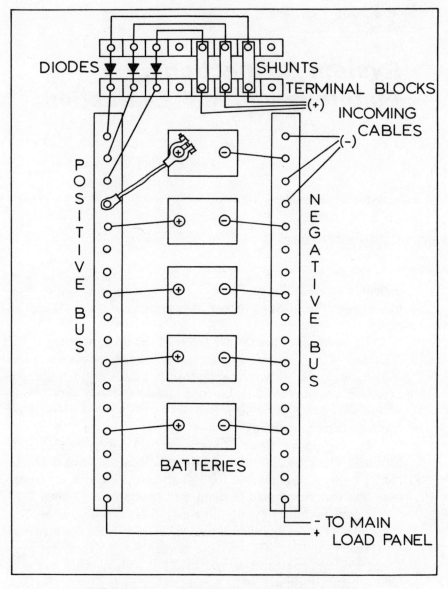

FIGURE 15-1. BATTERY AREA EQUIPMENT

along each bus, drill and tap holes that will accept the ¼-inch or ⁵/₁₆-inch × 1-inch bolts you selected. Use lockwashers under the bolt heads and do not overtighten to avoid stripping threads. The bus may be mounted on insulators of your choice in the battery box. Provide terminal strips with barriers between terminals for mounting diodes and current shunts. These may be mounted outside of the battery box on systems of less than 48 volts. On higher voltage systems, mount inside the battery box or other suitable cabinet to protect personnel from shock.

The battery box may be built in any fashion that will fit everything in. (See Figure 15-2 for a photo of my first battery box made from scrounged material.) The buses are made from two ½-inch copper pipes smashed flat with a hammer and bolted together with wiring sandwiched between copper pipe layers. This works but is cumbersome to use. If you put a lid on your battery box as I did, the box must have some large holes for ventilation (you need a hole for cable access anyway). Leave extra space for later battery additions and locate the buses so that the later batteries will have easy access also.

During the initial operation of your system, you may find battery discharge a problem until your system is balanced. This can be cured temporarily by charging your batteries with a battery charger. Connection can be made to the battery bus or to the main bus in the load panel for charging. Mark your buses positive or negative clearly so that no future connection errors occur. If for some reason you do not enclose your batteries, be sure to set them on a board and not on a concrete floor. Concrete has a way of ruining a battery chemically.

Some good anticorrosion discs are available at auto stores to go under battery connections. These will save you a lot of future work. Buy them and use them.

Battery maintenance is important even though very little is required. Remember, improper care of batteries will cause your whole system to fail!

CONNECTING LIGHTING AND OTHER LOADS

The electrical loads must be divided into two categories for discussion: direct current (DC) and alternating current (AC) loads. DC loads will be discussed first.

FIGURE 15-2. BATTERY BOX

All loads should be connected through a fuse or circuit breaker. In the case of DC loads, these will be located in the DC load panel. DC loads are categorized as incandescent lighting fixtures, fluorescent lighting fixtures, DC appliances, and inverters.

All DC loads will have a positive and negative wire to connect. The negative wire may go straight to the battery or come from the negative bus in the DC fuse or circuit breaker panel. The positive wire is connected to a fuse or circuit breaker. Sometimes more than one load is connected to the same fuse or circuit breaker. If this is done, be sure the fuse or circuit breaker, and wire are large enough for all loads that are connected to operate simultaneously.

Incandescent light fixtures (all those that are not fluorescent and light instantly) do not care whether the connection is AC or DC. For example, a standard house light bulb will run just as well on 125 VDC as it will on 120 VAC. But the voltage must be correct! A 12-volt incandescent light must run on 12 volts AC or DC, etc. Although incandescent lighting will operate regardless of which lead you connect to which, *always* connect the negative lead to the wire on the fixture that is connected (grounded) to the fixture housing. If you connect this lead to negative on one fixture and positive on another, the instant both are on at the same time, the battery will be shorted positive to negative. If you have used fuses or circuit breakers wisely, they will blow or pop. If not, you will have a wiring fire.

DC fluorescent fixtures have solid-state inverters built into them to create the AC that actually goes to the tubular lamp. You must observe the polarity given to you by the fixture manufacturer if this unit is to work. If no separate instructions are provided, look for a sticker on the inside where connections are made. Lacking this, ask your seller for instructions.

DC appliances are made specifically for DC applications. Do not connect any AC appliance to your DC system. The appliance must carry a nameplate reference to the exact voltage and designate DC before it is suitable on a DC system. Exceptions to this are items that have cigarette lighter plugs or are known to be auto parts. These may be connected to 12 VDC systems safely. On cigarette lighter plugs, the center terminal is positive.

Many portable TV's can operate on 120 VAC or 12 VDC. Follow the manufacturer's instructions in connecting these. If you have any doubt, ask your seller how to connect it.

Inverters are made for the specific purpose of changing DC to AC. Most available inverters convert 12 VDC to 120 VAC. So if you have an AC load that you want to run from your DC system, you have only to purchase an inverter and connect it between the battery and the AC load. When purchasing the inverter be sure to tell your seller what equipment you will be running with it. The wattage rating of the inverter must be higher than or equal to the AC load. If the output of the inverter is a square wave ask your seller if it will run your load. (See the last few pages of Chapter 8 for a discussion of sine wave and square wave loads that are normally acceptable.) Color televisions and refrigerators or freezers usually require a sine-wave AC power input. Televisions shouldn't be damaged by square wave input but they may be too distorted for good viewing.

AC loads can be connected in a normal fashion to AC outputs from inverters. The output terminals may read AC, AC and gnd, AC high (LH), AC neutral (LN), and have a chassis screw (maybe green) for ground. You should always connect three wires to AC loads: green or bare wire to gnd or chassis or fixture case; white wire to neutral, AC neutral, LN, or just one of the AC terminals if not otherwise marked; black wire to AC high, LH, or just one of the AC terminals if not otherwise marked. Sometimes the connections will only be marked by screw colors such as in a wall receptacle. If so, put the green or bare wire to the green screw, the white wire to the silver screw, and the black wire to the gold or brass screw. If your circuit breaker trips immediately on turning the device on the first time, go back to the fixtures you just installed on the circuit and make sure the bare ground wire isn't bent over onto one of the other terminals.

Should any problems or question arise on AC fixture connections consult your seller. Usually, he can help you.

When you add loads to your system, always add a new fuse or circuit breaker in your DC panel. If you connect directly to the battery bus you risk a wiring fire that could burn down your house.

Normally, no load problems are encountered with lighting fixtures. You may wish to modify a fixture, expecially one that has a switch built in. If you can't reach the switch, disconnect the switch and run the two wires to a wall switch. It doesn't matter which side of the switch you connect the two wires to. No ground wire is required on a switch. If you have more than two terminals you will have to figure out which ones to use.

HOW TO MAINTAIN YOUR SYSTEM

In this section, care for moving parts, batteries, terminals, and loads is discussed including troubleshooting procedures.

Moving parts are bearings, belts, chains, gears, etc. Chains and gears must have lubrication to prevent rapid wear. Chains may be lubricated by hand about once a week, but something like a wick oiler which gives continuous lubrication is better if you can work it into the design. Gears should have continuous splash lube if they are to last, however, many gears are grease-covered inside of a housing. If greased, check for grease once a year and whenever you see lube running out at seals.

Belts recommended in this book are all ½ inch wide auto belts. Just like in autos, they should have ½ inch side play (no more or no less). If the belt has less than ½ inch side play, it will excessively load the bearings resulting in eventual bearing failure. If the belt is too loose, it will slip causing loss of transmitted power and excessive belt wear. Never use a crow bar or similar tool to adjust a belt. It usually results in pulley or nearby housing nicks and cracks, and a belt that is too tight.

Bearings should have grease fittings built-in for periodic relubrication. Set screws on bearings to shaft should be torqued to 5.5 foot pounds (¼ inch screws) or hand tight if you don't have an allen-type torque wrench. I made all mine hand tight with no problem of looseness or overtightening. Your bearing should come lubricated but it is a good idea to check by relubricating it slowly while turning it until a bead of grease appears at the seals. (I use this method and found a new one that was almost dry.) Relubrication intervals are dependent on RPM, temperature, and cleanliness of use area. Figure 15-3 shows relubrication intervals for varying conditions, temperatures, and RPM. Use grease that conforms to NLGI No. 2 penetration.

Shafts of cold roll steel that are unpainted will rust. A preventa-

RPM	TEMPERATURE	CLEANLINESS	GREASING INTERVAL
100	up to 120°F	Clean	6 to 12 months
500	up to 150°F	Clean	2 to 6 months
1000	up to 210°F	Clean	2 wks to 2 months
1500	over 210°F	Clean	Weekly
any speed	up to 150°F	Dirty	1 wk to 1 month
any speed	over 150°F	Dirty	Daily to 2 weeks
any speed	any temperature	Very dirty	Daily to 2 weeks
any speed	any temperature	Extreme conditions	Daily to 2 weeks

FIGURE 15-3. REGREASING INTERVALS FOR BEARINGS

tive for this is to wipe silicon grease on the shaft at all exposed places. A thin film will do, but the more the better. Repeat about every 6 months to guarantee no rust. Shafts can be wiped with silicon grease prior to installation if desired. I find that this does not affect the ability of bearing set screws to hold.

Batteries are usually as good as their maintenance. Four areas on batteries require periodic attention. About once per month, check them all. They are dirty electrical connections, dirty top case, fluid level, and charge status.

When checking terminals for high impedance or corrosion, a visual look is insufficient. The only way to be sure that a terminal won't give you trouble tomorrow is to loosen it and wire brush the post and connector hole. Do them all once a month and you will never have any problems here. Since many terminals are in close proximity to each other, extreme care to prevent touching a positive lead to a negative connection is worthwhile. If this happens, severe arcing will occur, and you could be burned. The bus connections need not be cleaned—just those at the batteries. If your climate is dry, you may get by cleaning the battery terminals once every 6 months—no guarantee.

Each month wipe the top of the battery cases off to prevent leakage currents from flowing. A clean, dry rag is best. If you really have a mess to clean off, try baking soda solution. But if you use baking soda, be sure that none enters the cells as it will negate the acid.

Each month, check the fluid levels on each cell of the batteries and fill them with distilled water to the ring or other manufacturer's mark. Using tap water will reduce the life of the battery due to miscellaneous mineral deposits in impure water.

Each month, check the specific gravity of the acid solution in each cell. This will give you the charge status of your batteries. All batteries connected in parallel should have the same cell readings. If not, you have a problem. If one cell reads discharged on a 6-cell battery, it is shorted and battery replacement is in order. If they all read low, then either that battery has a dirty terminal or insufficient charging power is being delivered (perhaps due to lack of wind at the wind turbine). When a lead-acid battery reaches the end of its life without an individual cell going bad, all cells in the battery will be just into the red on the hydrometer and continue to get worse. Internal current leakage will prevent this battery from ever reaching a normal charge. If in parallel with other batteries it will bring down the whole battery group and it will be difficult to locate the problem. Replace it, when found. Specific gravity is checked with a hydrometer. These can be obtained at your auto store. They usually have green, white, and red areas on the float: red being discharged to some level between 1.100 and 1.224 specific gravity; white means partially discharged and still OK between 1.225 and 1.254 specific gravity; full charge is indicated by green between 1.255 and 1.300 specific gravity. You should always take the reading before adding water as you will get an erroneously low reading just after adding water.

When troubleshooting loads not receiving power, first check fuses or circuit breakers. Second, check that switches are in the correct position. Third, be sure the battery is connected. If all these fail to yield the cause of no power, see if the light bulb is burned out (for lights). Barring that, get out a voltmeter, start at the battery and work toward the load until voltage is lost. Then check for a broken wire or loose connection at the lost-voltage location.

Motors and generators are usually permanetly lubed, but most have brushes that wear out. Annually, check your brushes for wear. Otherwise, check them when proper RPM's do not respond with proper output voltages. If your motor burns up (wiring) or you don't know how to replace the brushes, contact a motor rewinding firm. These firms are available in almost all towns of any significant size.

They are usually real experts and can have your motor or generator back in very short time as good as new. This is much cheaper than buying a new one. A word of caution on permanent magnet field motors: Don't take them apart just to look. Every separation of armature from field reduces the magnetic flux in the field. It can be rejuvinated, but not without special equipment.

For unusual problems, apply common sense or ask someone that sells the unit you are having trouble with about its maintenance.

EVALUATING CAPABILITY, EFFICIENCY, AND EXPANSION

In this last section, let's take a hard look at what we have accomplished. First, you will want to know how many kilowatt hours or what peak power you can attain from your generator units. Since watt meters and kilowatt-hour meters are not in everybody's closet, I expect you to make these measurements by hand from simultaneous ammeter and voltmeter readings. Get someone to help you, if you need an extra set of eyes.

On each generator unit in your system take a number of readings of volts and amps out at the generator or before it passes the diode. Multiplying amps times volts will give you watts. Divide by 1000 and you have kilowatts. Estimate over how many hours you can average that output in a month, multiply those hours by the kilowatts and you have kilowatt-hours. Multiply the kilowatt-hours by the electric rate and you have cost savings per month. Don't expect to see a lot of money since 3 to 5 cents per KWH doesn't add up very fast.

Do these readings on all your generator units. If they are wind generators, get the wind speed at the same time if you can. This requires a wind speed indicator of some commercial type. If it is a water wheel, then either know the water flow and head or measure them again.

Efficiency can be calculated for any generator if you know the input power and output power for the same instant. Efficiency varies with power input and a curve can be plotted of efficiency VS input power.

$$\% \text{ Efficiency} = \frac{\text{output power (100)}}{\text{input power}}$$

Output power in a DC generation system is always volts times amps. Input power is calculated for wind as shown in Chapter 4, for water as shown in Chapter 6, and for sun as shown in Chapter 5. For water power, convert to watts by making 1 HP = 746 watts. For sun, you are working with an approximation of peak power, but see Chapter 14 for examples of efficiency. The efficiencies at medium loads should fall bewteen 20 and 30% for vertical wind turbines; 25 and 50% for propeller wind turbines; 25 and 60% for water wheels; 10% for photovoltaic silicon cells; and about 25% for heat-to-motion sun usage. These efficiencies will tell you how good a designer and builder you are. Don't expect miracles though; high efficiencies are a result of breakthroughs in friction and other losses.

Compare your electric bills and kilowatt-hour estimates on your equipment. Subtracting one month from the next doesn't tell you much. You will have to compare last year's receipts to this year's for the same month taking into account rate changes and weather differences. At best, it's an educated guess of what you really save. But this should give you a general idea.

A method for determining the annual operational costs of small electric plants is recommended in the *Standard Handbook for Electrical Engineers*, ninth edition as follows:

Assuming the life of the wind, water, or sun generator system to be 10 years and the storage batteries as 6 years, calculate your annual costs as follows:

ANNUAL FIXED CHARGES:
Interest on investment at 6%, on half valuation.
Depreciation on wind, water, or sun generator unit at 10%.
Depreciation on storage batteries at 16⅔%.
Depreciation on wiring at 4%.
Taxes at 2%, on half valuation.
Insurance at actual cost, on half valuation.
Operating Expenses: maintenance including labor at actual cost.

The largest costs are depreciation on batteries, depreciation on generator system, and interest on investment, in that order. The total of the above list will tell you what the average annual cost will be to keep your system operational indefinitely; replacement cost is included in depreciation.

You may consider expansion of your system. If you are only

doing so to save money, I recommend that you check your cost estimate closely to see if it's worthwhile. Estimate your energy savings from the expected added power. Then compare this to the initial cost and expected maintenance.

For me, reliability and control of loads are important considerations. I just picked up an issue of "Pacific Gas and Electric (PG&E) Progress," a monthly consumer information sheet, and what I read scared me. I quote: "During the record-breaking heat at the end of June, all available generating resources in Northern and Central California were in use. Demands cut deeply into reserve capacity, and all available power from outside the area was brought in from both North and South—the Pacific Intertie was fully loaded with power from the Pacific Northwest. The situation was one of the tightest the company has experienced in recent years. Had higher loads been experienced, or had there been a major equipment failure, rolling brownouts (limited service interruptions) might have become necessary."

In that same issue another article noted that should an electric shortage occur, the first things consumers would have to turn off would be swimming pool pumps, heating and cooling of unused spaces, ornamental lighting and displays, etc. In still another article, PG&E said they plan to test devices in 400 homes "that automatically disconnect residential air conditioners and electric water heaters during periods of high demand." But there is more, a rate increase has just been granted by the Public Utilities Commission for PG&E to offset their increased costs. My personal plans are to build a solar home, thereby cutting my electric needs in half. Then, I'm going to install five windmills and a mirror-steam system (designs from this book) to provide 100% of my electrical needs. The generators will charge 600 ampere-hours of batteries at 125 volts DC. The system will use 6 KW's of inverters to provide 117 VAC single phase for the entire house. I will keep the local power company tied in as automatic backup. I believe in this project although it will take about three years to finance the changes and not everyone will think it is worth the cost, but my cooling system and hot water heater won't be going off during peak demand periods!

I have enjoyed these experiments immensely and will continue to plan bigger and better projects for my personal use. I urge you to join the elite group of technicians and experimenters now enjoying solar power.

Formulas

WIND (ref. Chapter 4)

$$\text{Power in wind} = \mathbf{P_{in}} = \mathbf{K}\; \tfrac{1}{2}\; \rho\; \mathbf{AV^3};$$

where P = power in watts,

ρ = density of air in slugs/cu.ft.,

A = area (sihouette) or rotating components in sq. ft.,

V = speed of wind in ft./sec. (MPH $\times$ 1.47 = ft/sec.),

K = 1.356 (conversion of ft. lbs./sec. to watts).

The value of ρ is .002377 at sea level (see the table in Chapter 4 for other altitudes). Chapter 4 has example applications of the wind equation.)

$$\text{Power output from windmill} = \mathbf{P_{out}} = \mathbf{IV};$$

where V = volts and

I = current in amperes, and

P = power in watts.

$$\text{Windmill efficency (\%)} = E = \frac{\mathbf{P_{out}}}{\mathbf{P_{in}}}\; (100);$$

where E = efficiency in percent, P_{out} is as above, P_{in} is as above, and 100 is the percent conversion factor.

It should be noted that this efficiency will only apply at the wind speed used in the P_{in} equation. Efficiency will form a curve with increasing efficiency as wind speed increases to a certain point where a level off of efficiency occurs.

SUN (ref. Chapter 5)

$$\text{Solar Constant} + 1353 \text{ watts/M}^2 =$$

$$428 \text{ BTU/ft}^2\text{hr} = 125.7 \text{ watts/ft}^2$$

(See Chapter 5 for application of solar constant.)

PHOTOVOLTAIC CELL GROUP CHARGING BATTERY

$$I_{charge} = \frac{(V_{oc} - V_B) I_{sc}}{V_{oc}}$$

where I_{charge} is the battery charging current expected,

V_{oc} = total open circuit voltage of the photovoltaic cell group in sunlight,

V_B = battery voltage including the series diode voltage drop,

I_{sc} = short circuit current of the photovoltaic cell group in sunlight.

$$\text{Efficiency of photovoltaic cells (\%)} = \frac{I \, V \, (100)}{A \, K} \, ,$$

where

I = milliamperes output while loaded in noonday sunlight,

V = volts output while loaded in noonday sunlight,

A = area in square centimeters,

K = 100 milliwatts per square centimeter.

Cells should be normal to sun during measurements. This is not a laboratory condition but will suffice for experimenters. In laboratory conditions, measure the light source in milliwatts per square centimeter and substitute for K the measured amount.

WATER (ref. Chapter 6)

Available horsepower input from water $=$ $HP_{in} = \dfrac{W\ Q\ H}{33,000}$,

where HP_{in} = horsepower in,
  W = 62.42 pounds per cu.ft (water at 40°F),
  Q = water flow in cu.ft./minute,
  H = head in feet (difference in height of entering and exiting
        water to water wheel), and
1/33,000 = horsepower conversion factor.

Water wheel efficiency (%) $=$ $E = \dfrac{VI\ (100)}{HP_{in}\ (746)}$

where E = efficiency of water wheel in percent,
HP_{in} is as above equation,
746 is the conversion factor for watts per horsepower,
100 is the percent conversion factor, and
VI is measured power output with V in volts and I in amperes.
50% can be reached for overshot water wheels, other types will be
  less efficient.

Water wheel width $=$ $W = \dfrac{Flow \times 3}{(RPM)(No.\ of\ buckets)(Section\ of\ bucket)}$

where W is in feet,
3 is for ⅓ full buckets,
RPM is revolutions per minute from Figure 12-1,
No. of buckets (usually 8) but could be more on large wheel,
Section of bucket = ½ depth × width (if triangular) in sq. ft.

BIBLIOGRAPHY

Clegg, Peter, *New Low-Cost Sources of Energy for the Home*. Charlotte, Vermont: Garden Way Publishing, 1975. Includes information on solar energy, wind power, water power, water/waste systems, and wood heating. Each section includes a catalog of suppliers.

Daniels, Farrington, *Direct Use of the Sun's Energy*. New York, N. Y.: Ballantine Books, 1974. Includes information on collectors, cooking, heating, distillation, storage, drying, cooling, heat engines, electrical and chemical conversion, and other uses of the sun.

Daniels, George Emery, *Solar Homes and Sun Heating*. San Francisco, Calif.: Harper & Row, Publishers, 1976. Tells how to use the sun to heat homes. Includes information on heat losses useful in mirror heat applications.

Duffie, John A., and William A. Beckman, *Solar Energy Thermal Processes*. New York, N.Y.: John Wiley & Sons, 1974. Summarizes the state of knowledge useful to engineers in understanding and designing solar processes. Includes theory on solar radiation, heat transfer, material radiation and transmission, collectors, storage, solar hot water, and solar heating and cooling.

Hackleman, Michael A., and David W. House, *Wind & Windspinners*. Culver City, California: Peace Press, 1974. Tells how to build the S-rotor Savonius windmill.

Wade, Gary, *Homegrown Energy-Power for the Home and Homestead*. Willits, California: Oliver Press, 1974. A list of companies and what they make in solar energy products.

INDEX